TOXIC BEDLAM & MISADVENTURE

BY

VINSON CHARD

Published by New Generation Publishing in 2018

Copyright © Phil Jones 2018

First Edition

The author asserts the moral right under the Copyright, Designs and Patents Act 1988 to be identified as the author of this work.

All Rights reserved. No part of this publication may be reproduced, stored in a retrieval system or transmitted, in any form or by any means without the prior consent of the author, nor be otherwise circulated in any form of binding or cover other than that which it is published and without a similar condition being imposed on the subsequent purchaser.

www.newgeneration-publishing.com

 **New Generation Publishing**

CHAPTER 1

The job advertised on behalf of a company, called Chemkiln, and displayed in the employment section of a national newspaper during the mid-nineties, virtually jumped out at me, while at the same time shouting, 'Apply, apply!' I had to agree with my inner voice, the position did appear to have my name attached to it. For the advert requested experience of certain tasks and duties which mirrored those I performed on behalf of Hyperwaste, my employer at the time.

Supervising Project Manager(s) Required Chemkiln

Personnel required with Chemical background, ideally Chemist/Chemical Engineer for a multi-national company specialising in high-temperature incineration and the disposal of miscellaneous toxic chemical wastes. The position comes with a salary in excess of £26K and negotiable dependent upon experience. The package also includes a new company car, expense account, overtime payments and a generous company pension scheme. Applicants must have experience in the labelling and shipping of hazardous material and exhibit a willingness to travel extensively throughout the UK, and sometimes abroad. Applicants must also have a 'hands on' approach to large waste projects and be willing to assist and lead by example. If you feel this position is for you, please contact etc., etc.

By this time in my career, I began sensing my incumbent employer Hyperwaste, appeared distinctly more than reluctant in paying me the remuneration which I felt morally and justifiably entitled to, considering all the time and effort being devoted to the job on my part. It seemed Nathan Edwards, one of the more senior directors and my immediate boss, was not playing fair and, putting it bluntly, simply taking the piss. The unfortunate aspect of this was that I got on extremely well with Nathan; I liked and respected him both as a boss and as a person. However, when it came to money and pay awards, he did have the annoying tendency of being as tight as a duck's arse. Despite my position within the company having a dotted line of responsibility to a less senior director named Roger Tate, when it came to the purse strings, Nathan maintained the authority, and, indeed, he held those very strings tightly clasped in his unyielding hands.

Circumstances had altered from a few years earlier, when, during his time as Area Sales Director, my line Manager at that instance, Dave Pearce, had been good enough to give me a substantial increase in salary, awarding pay increases annually. Also, he tended to be extremely generous with those annual settlements. Unfortunately, my former munificent benefactor had now long since gone. Following his departure, my remuneration subsequently stagnated. Throughout the ensuing three years with Hyperwaste, there had been no real pay increase worth mentioning. Whenever I plucked up the courage to broach Nathan Edwards, vis-à-vis a salary increase, I generally obtained some feeble excuse about getting the newly acquired treatment plant fully operational and making it a viable, profitable enterprise before any more pay increases could be awarded, even suggesting I could be given an increase in salary, possibly based on profit sharing in the distant future. Nathan dangled the proverbial carrot in front of me, treating me like an easily manipulated donkey. During my student days, one of the college lecturers in the department often preached and pontificated the following:

'After about six or seven years, you have nothing more to offer a company, and they have nothing more to offer you. It is then time to seek another job and leave.'

Probably complete and utter bullshit. But coincidentally, seven years had indeed passed since launching my career with Hyperwaste, and I began considering it time to re-shuffle my personal deck of cards, move on, and, hopefully, broaden my horizons along the way, with the possibility of better prospects bestowing themselves upon me during the journey. Nevertheless, a substantial increase in salary by Hyperwaste might have altered and completely reversed my stance on such an opinion.

Also, I became more than slightly disgruntled and disillusioned with the hours I worked and put in, mainly helping set up and commission the newly modified oil treatment plant acquired from the tricky, duplicitous, narcissistic Sir Andrew Chadrock Hulsey, the Donald Trump of the waste industry. I also had extra responsibility thrust upon me, yet not receiving the just rewards. My old school motto of *post luctum, fructum*, (after toil, fruit) just did not seem to apply, indeed, it appeared to be more a case of; *post luctum, luctum,*(after toil, {*more*}toil).

So that eventful day when I accidentally stumbled upon the advert in the newspaper for a General Services Supervisor with Chemkiln, a well-known toxic waste disposal company specialising in the high-temperature incineration of dangerous chemicals and based in South Wales, the opportunity presented itself as manna from heaven. I became completely captivated by the advert. It would mean a salary increase of thirty percent, a new company car, expenses, and, more importantly, overtime payment. The job involved sorting hazardous, toxic waste for incineration and safely packing it ready for transportation, ensuring the correct procedures were adhered to for shipment of the waste by road. Just the type of work in which I had accumulated a vast amount of experience, plus being fully conversant with the intensifying hazardous waste laws and regulations, the job seemed to be just up my street. I perceived the only hurdle and major obstacle to overcome in obtaining the position, being my advancing age. By now, I had reached my early forties, with ageism rife in the workplace, not an ideal period in one's life for changing jobs. I considered the notion if I didn't apply immediately it could be one of those missed opportunities experienced throughout life, generating the proverbial '*what if*' scenario. So forthwith, without any more hesitation, I popped my written job application along with my updated, bespoke curriculum vitae, into the nearest post box.

Following an interval of a couple of weeks, I received a written reply, inviting me to attend an interview at the huge incineration facility, located just a few miles from my home town, where I now lived with Stella. I accepted the invitation without any form of hesitancy or reticence on my part.

The company scheduled the interview for lunchtime on a Friday. I asked for the afternoon to be taken as part of my holiday entitlement, but unfortunately, experienced great difficulty in getting away. For that particular day, Nathan Edwards decided to pop into the treatment plant at Newport, on one of his all too frequent visits. With him being such a garrulous and loquacious person, I experienced great difficulty in making an escape from my workplace, allowing me ample time to change into the obligatory interview suit somewhere en route. Finally, I eventually managed to take flight and, along the way, quickly change into my interview attire before finally arriving at the Chemkiln facility, only a mere five minutes late. The hair-raising, swift journey required me to exceed all the legally enforced speed limits. Fortunately, the interview with the previous candidate extended well over the schedule. The only person cognizant of my late arrival was the receptionist, hence, my lateness went completely unobserved by my interrogators and without causing any upheaval to the interviewing schedule, negating any possible annoyance to the interviewers and probably generating a black mark against me.

The time arrived for my interview. It may be coincidental, but a fair number of my job interviews managed to encompass something special and memorable about them. This particular interview with my interrogators, Rupert St. John Smythe and Albert Parry, two senior members of the General Services department, proved to be no exception, succeeding in becoming quite bizarre and surreal. The two interviewers had obviously been allocated with the task of filtering and sifting out the most suitable candidates before making their recommendation to the head of department, Dr. Damien Phipps. He would then ultimately make the final decision concerning who to employ.

Being extremely tall, Rupert towered above me at approximately 6 ft 7 inches in height. As well as being exceedingly thin, and bespectacled, Rupert sported a mop of hair, which appeared to enjoy a life of its own. The strands of his unkempt mop protruded in all sorts of independent, miscellaneous directions. I guessed Rupert to be in his mid-thirties. He also spoke with a very refined, well-educated, softly spoken, eccentric, English accent. Obviously English born, originating on the Anglo-Saxon side of Offa's Dyke. Although I didn't hold that against him, as he seemed to be a pleasant enough fellow. Because of his appearance, idiosyncratic behaviour and somewhat confused, befuddled demeanour, he reminded me of a quasi, thin, tall, bespectacled Boris Johnson.

Albert, on the other hand, measured in about average height, displaying an impressive beer pot, with a head as bald as a newly hatched egg. He also wore spectacles. I estimated this particular individual to be somewhere in his late-fifties, or possibly early sixties. However, in complete contrast to Rupert's jovial, smiling deportment; Albert possessed a stern, surly and distinctly, curmudgeonly, grim demeanour. Like me, he spoke with a discernible South Wales valley's accent.

As previously alluded to, the interview became completely bizarre and surreal. I had a couple of years earlier, obtained an 'A' Level in Film studies, a hobby and obsession of mine. Throughout my youth, my father enjoyed taking me to the cinema, or pictures, as they were then referred to in those days gone by. Introducing me to film classics such as; *The Magnificent Seven*, *The Guns of Navarone*, *Enemy Below*, *The Great Escape*, *The Longest Day*, *The Bulldog Breed*, *The Fast Lady* and *How the West Was Won;* to mention just a few. Through this bonding process with my father, I became enamoured by the movies, thoroughly enjoying them. In my later life, and because of this infatuation with the film business, I embarked upon attending evening classes at the local college, with the prime objective of acquiring more insight concerning the film industry which so captivated me. Rupert St. John Smythe, I quickly discovered, also proved to be a film buff. He appeared to be completely intrigued and fascinated with the film course after reading about my interest in the subject, which he discovered tucked away in the hobbies and activities section of my curriculum vitae. My inquisitor exhibited so much interest, he apportioned a considerable amount of interview time to discussing the film course and films in general; not asking one technical question relating to chemistry, chemical engineering or the intricacies of the numerous laws concerning waste management. Instead of discussing thermodynamics, along with the theory of high temperature incineration, and discussing other technicalities about the chemicals which were prohibited from being vaporised in the kiln, or the safety aspects and considerations when transporting miscellaneous toxic chemicals, we ended up discussing the influences exerted by the German Expressionist cinema made in the twenties upon the subsequent American Film Noir of the forties and fifties, as well as the films of the great German film director, Fritz Lang. Rupert and I also talked about the minutiae, subtext and nuance of Alfred Hitchcock films, together with the concept of the auteur film director. We also chatted briefly about the theory of montage and juxtaposition extolled by the great Russian film director Sergei Eisenstein. All these topics, a million miles from chemistry and engineering. Albert soon became utterly bored with the whole situation, and the seemingly never-ending discussion taking place between Rupert and myself. Consequently, he contributed very little to the proceedings, only asking one solitary question concerning my drinking habits and whether I drank alcohol. I deliberated the query extremely carefully before answering. After all, this could be a trick question as I considered the possibility of Albert being some sort of white Evangelical or temperance zealot. His motivation in asking that question to ultimately berate me upon discovering I consumed alcohol, classifying me as an un-Godly Heathen. Alternatively, did he simply want to know if I was the sociable type? Following careful deliberation on my part and, after considering Albert's sizeable paunch, I decided to be utterly honest and truthful about my drinking habits.

'Yeah, sure, I drink alcohol, not to excess, but I enjoy having a few pints and socialising whenever the opportunity presents itself,' I answered straightforwardly, while at the same time, looking Albert directly in the eye. Only then, did I detect a slight, barely perceptible smile on his face and what I considered to be an indication of his approbation and approval concerning my reply, realising my answer had hit the correct spot. Consequently, I would not be subjected to a discourse and berating on the dangers of the demon drink, with the prospect of eternal hellfire and damnation in the after-life. Rupert posed a few additional minor questions concerning my experience with the packaging of chemicals safely for shipment by road, and that was the end of the interview.

I could not believe there had not been any in-depth technical discussions concerning the intricacies of high-temperature incineration, together with questions regarding calorific values, thermodynamics or the types of chemicals suitable or unsuitable for the incineration process as a means of waste disposal. I later discovered the General Services team had already made discrete, surreptitious enquiries, quizzing Hyperwaste personnel as to my technical competence and capabilities, evidently approving of what they gleaned from the individuals being questioned. The interview process' primary objective was to ascertain whether I would slot into the team and be able to get along fairly easily with other members of the department.

A further week or so passed by before Chemkiln made contact again. While at work, I received, a telephone call from Rupert St. John Smythe asking if I wished to attend a second interview, this time with Dr. Damien Phipps, head of the General Services department. Once again, I agreed to be present at the arranged interview. Well, at least, I had made it to phase two of the process. It appeared I must have come across reasonably well during my first interview.

The day of the second meeting arrived. This time, ensuring there would be no repetition of the previous interview, I booked the day off work in preparation. I arrived slightly early on the site and sat composed, not sweating or agitated as I had been before the previous interview. Damien Phipps came out of his office and invited me in for the second part of the elimination process. He proved to be nothing like I had imagined, attired in cargo trousers festooned with a multitude of pockets and wearing a navy-blue tee shirt bearing the Chemkiln logo on his right breast. He was heftily built, I estimated him to be approximately 5 ft. 8 inches tall. I guessed him to be in his mid to late thirties. However, he did not have the general appearance of the stereotypical academic, with the accoutrements of dishevelled hair, thick spectacles, wearing a threadbare jacket, displaying leather patches on the elbows. These images being the vision I had formulated in my head prior to our meeting.

Upon his entering the room, the interview began in earnest. I knew this time the questions would be more searching and in-depth. He asked me to explain my duties at Hyperwaste, requesting a brief job description. I went into detail about my duties and what was required of me. He then asked about my experience in preparing chemicals for shipping by road and if I was *au fait* with the IMDG (International Maritime Dangerous Goods) regulations for shipping waste and chemicals by sea. At that time, I had no experience of the IMDG regulations; informing him of the fact.

He told me it did not matter as he had scheduled all the supervisors in the General Services department to attend an intensive IMDG course in the near future. He then asked me a question which I somehow felt prepared for.

'Would you consider yourself to be a proactive or reactive person?'

My answer seemed to effuse from my mouth before even thinking about it.

'Both,' I replied, before confidently going on to explain.

'There are instances of both, one must prepare for projects, but there are also instances when unforeseen circumstances arise. That is when a person must be reactive.'

Dr. Phipps looked at me thoughtfully.

'I'll say this; you are certainly good at being interviewed and you have a polished interview technique' he commented, while at the same time, wagging his right index finger at me, generating more than a slight hint of sarcasm in his voice.

'Not really,' I replied. 'Just answering as I believe I generally behave.'

The interview went along with the usual interview profiling. Dr. Phipps must have just attended an interview technique course; all his questions appeared to be straight out of the text book, '*Interviewing Techniques for Dummies.*' Questions such as what I considered to be my strengths, my weaknesses, tell me a bit about yourself. What makes you tick?'

I had, by this stage in my life attended so many interviews, Damien proved to be spot on in that respect, I did have a vast experience of interviews. I began telling him about my interests, how I worked, believing in leading by example and my time with Hyperwaste. He did not ask too many technical aspects in the interview and after an hour it concluded with me asking the usual questions about pay, hours of work, holidays and pension schemes etc.

We shook hands, Damien then explained he had a few more candidates to interview and would let me know in the next couple of weeks, whatever the outcome of my application. I believed I stood no chance due to my age; but had enjoyed the experience of the two interviews. They had been interesting to say the least.

A week or so later while at work, I received a telephone call from Dr. Phipps. After first enquiring if I could talk freely, he informed me I had indeed been successful in getting the job, before going through the salary and conditions on offer. He also told me I would be receiving the official acceptance early the following week, with the proviso I still wanted the position. After being informed of the salary on offer, there was absolutely no doubt in my mind. I would be most certainly accepting the offer and told Dr. Phipps without any form of ambiguity and possible misunderstanding. He also asked if I could still attend the IMDG course booked for a fortnight's time and which would last for three days. I informed him it would not be a problem.

The following Monday, Nathan Edwards visited the facility. The moment he entered the building, I called him aside and handed him my formally typed, required, four weeks termination notice. Nathan appeared taken aback; he had not anticipated this sequence of events. Upon interrogation, I told him it was purely the remuneration on offer and that Hyperwaste had kept me dangling on a string with pay awards. This pay increase with the new company was definite, with the possibility of extensive travel. When I told Nathan the actual salary on offer, he said he could not match it. So, with some reluctance, he accepted my resignation.

Roger Tate, the other director came to see me a few days later, telling me he wished I had approached him earlier about my salary rather than approach Nathan. He told me he might have been able to do something about it.

Before the final termination of my employment with Hyperwaste, I booked a few days of my holiday entitlement to attend the IMDG course being held in London and organised by PIRA. I met the other members of the General Services department and we drove up in a convoy of company cars. It was then I met the other supervisors and other members of the department. I enjoyed their company and believed I had made the correct decision in accepting the position and the prospect of a new career direction.

The team quizzed me about Hyperwaste, specifically, the newly acquired treatment plant. Of course, the inevitable topic of Sir Andrew Hulsey came up during the discussion; for the General Services team had also dealt with him, with one of the previous managers, Jim Beam, becoming very friendly with the duplicitous Baronet. But as Damien sagely pointed out, having a thorough knowledge and contact with the tricky aristocrat, Sir Andrew Hulsey only befriended someone if there was some mileage or benefit in it for him. In the case of Jim Beam, Sir Andrew had the loan of a quantity of specialised, extremely expensive chemical suits, breathing masks and oxygen cylinders. None of which were ever returned to Chemkiln. Sir Andrew Hulsey denying vehemently he ever had them in the first place. The cost to Chemkiln of replacing the equipment ran into thousands of pounds. Typical Sir Andrew Hulsey, always on *'the make.'* The General Services team evidently also had no truck with, or time for the mendacious, duplicitous Baronet.

The IMDG course turned out to be quite gruelling and tough, with a large amount of concentration required, enabling me to pass the stringent examination at the end. Thankfully, together with all my new colleagues, I passed the course.

The final weeks with Hyperwaste passed quickly, while I tried to pass on all relevant information on to Paul, my assistant, as well as Nathan who would assume responsibility for the plant following my departure.

On my final day working for Hyperwaste, I booked a hotel room in Newport. The team from the treatment plant insisted on giving me a good night, accompanied by other staff from other Hyperwaste facilities, touring the flesh pots of Newport. A good evening was had by all, especially me. The night passing as a drunken haze.

However, I do vaguely recollect performing an impromptu striptease on the small stage of a local nightclub, instigated by the local DJ, and being egged on enthusiastically by my colleagues.

I did experience a slight tinge of sadness at leaving Hyperwaste and my work colleagues, believing this would be my last opportunity at a lucrative career change. But, I also looked forward to another new challenge in my fluid, ever-changing life.

CHAPTER 2

Throughout my career, I always somehow managed to gain employment with companies generating a high media profile and always in the news spotlight for one reason or another. Invariably, the media slant tended to lean towards the adverse or negative aspects of those companies. Obviously, the main reason for finding myself in this situation being, most of the firms I obtained employment with, in the manufacturing sector, tended to have a large involvement with chemicals and the chemical industry. An industry which is frequently portrayed as the scourge and pariah of the modern civilized world.

Regrettably, as far as dealing with chemicals is concerned, it appears to be the very nature of the beast, often generating into an emotive, political subject. Although, I must point out, society does tend to be somewhat hypocritical in its castigation of the manufacturing industry, with modern homosapiens possessing an insatiable, ever increasing desire to possess luxury items such as; mobile phones, tablets, computers, cars, domestic appliances and such like. All manufactured with the aid of chemicals. Unfortunately, because of its *raison d' être*, primarily, its involvement in the hazardous waste sector, Chemkiln also fell into the category of being a canker on humanity, particularly when it came to the media's perspective of the company.

Chemkiln began its life with a few enthusiastic chemists and engineers carrying out experiments by incinerating miscellaneous chemicals in small pilot plants. The ultimate objective, to dispose of and completely destroy most hazardous toxic chemical waste by means of high-temperature incineration, reaching furnace temperatures well in excess of 1,200 degrees Celsius, doing away with the need for landfill, and not leaving a ticking time bomb of buried, untreated, toxic chemicals for future generations: fairly laudable objectives really. The scientists took their ideas and theories to a large multi-national company who jumped at the concept; immediately building three high temperature incinerators: one in Falkirk, Scotland, one on the south coast of England and the last in South Wales. This third facility is the one where I later became based during my time with Chemkiln.

The incinerator in Scotland won the race, becoming the first to be commissioned. Unfortunately, it also suffered several major technical issues and was forced to shut down after a couple of years because of those very problems. The two other incinerator facilities ultimately profited from those technical difficulties and teething problems experienced in Scotland, finally surmounting most of the operational obstacles. The two remaining facilities operated for a while, but unfortunately failed to make a substantial profit. Most years, the finances tended to just break even. After several years the inevitable happened and both the incineration facilities began operating fiscally in the red. By this juncture, the board of the large holding company had had enough, deciding to sell off the two incinerators while they still could.

However, two of the directors on the board decided to try their luck on a management buy-out, offering to purchase the two facilities. Most of the other directors were only too glad ridding themselves of this aspect of their business which had been nothing but a thorn in their side from the outset. With huge running costs, technically difficult to operate, along with all the vitriolic, adverse publicity continually being heaped upon them. Plus, for no apparent reason, always the distinct possibility of the intermittent demonstration, which could suddenly erupt at the drop of a hat. In addition, some delinquent residents outside the perimeter fence, using air rifles, regularly taking pot shots at the company personnel working inside.

Because of all these difficulties and the desired quick sale, the two directors managed to buy the two sites for a measly two million pounds. The large company being keen to rid itself of the business as quickly as possible. It did not matter to the holding company that the price on offer proved to be at a minimum, '*be rid and be damned*,' seemed to be their philosophy. As luck would have it, the two purchasing directors bought the two facilities at the right time.

Business success is irrefutably built on a mixture of business acumen, knowledge of the product and ultimately, a fair proportion of generous assistance from Lady Luck. Around this time, fortuitously for the new owners, the worldwide electrical industry began divesting itself of the vast amounts of unwanted PCB (Polychlorinated Bi Phenyls), the synthetic oil used in the cooling of transformers. At the time, the only way environmentally acceptable of destroying the man-made oil, meant using high temperature incineration. The new owners had bought two of the only three high-temperature incinerators located in the UK with the capability of performing this operation. All for a bargain basement price at that.

Within a year or two of purchasing the two plants and researching the business, the new owners realized the company had been underselling itself and disposing of highly toxic wastes far more cheaply than similar operations based in Europe and throughout the rest of the world. They subsequently increased the price charged for high temperature incineration and disposal of the toxic chemicals, especially PCB. Being non-flammable, with no calorific value, the PCB transformer oil needed to be blended with combustible liquids such as flammable solvents and normal mineral oils to perform the required combustion process. Much to the horror and consternation of the chemical and waste industries, the new owners also increased the price for disposal of these chemicals. Although, the increase in the disposal price for those flammable chemicals tended to be nowhere near as expensive as the hugely increased price being charged for PCB and pesticides.

The customers disposing of PCB had no option but to agree to the new price hike. Within two years of their acquisitions, the new owners began making hefty profits largely because the volume of transformer oil being disposed of increased almost tenfold. They had purchased the two incinerators at the right time, with Lady Luck most definitely bestowing her favours on them.

It is rumoured Richard Branson began the same way with his Virgin Empire. His first company, the young, fledgling Virgin Records kick started his conglomerate, initially by producing and selling a fabulous vinyl album called Tubular Bells, written and performed by Mike Oldfield, the vinyl album being the catalyst in Mr. Branson's meteoric rise. For, at almost the same time of Tubular Bells' release in the early 70's, William Friedken, the director of the film 'The Exorcist,' decided to use some of the music from Tubular Bells for the sound track of his film. Consequently, sales of the iconic album rocketed, helping kick-start and ultimately establish the extensive Virgin Empire on the back of those massive vinyl sales. As the saying goes, the rest is history. So even the illustrious Mr. Branson had a smidgen of luck at the beginning of his entrepreneurial career.

Back to my story concerning Chemkiln. For the next few years the two plants made vast profits and the prospects for the business turned from being tenuous to one of becoming extremely secure and highly lucrative. The two directors would often turn up separately at the gates of the facilities, one in his Bentley, the other in his Rolls Royce. Each vehicle fully laden with gifts for their workforce. The handouts, generally consisted of extremely expensive, designer named perfumes for the females and Polaroid cameras, or such like, for the men. The reader must remember this all transpired prior to the advent and age of the digital camera, with Polaroid cameras being a highly-praised item at the time. The two directors would then quite happily distribute all these goodies, like benevolent, generous, beatific parents distributing Christmas gifts to their offspring.

Bernard Evans, one of the operators and colleagues in the General Services department, often enjoyed relating the story of his time as a union shop steward all those years earlier, way before my time with Chemkiln, when the two directors still owned the company. His eyes frequently glazed over, almost filling with tears, as he informed me of how he often participated in the yearly wage negotiations. The company had been owned by the new directors for a few years. During that time, it became evident to all concerned, their company seemed to be making money hand over fist.

The sheer increased volume of PCB waste material being disposed of by the company and the prices being charged for their disposal helped instil this belief. This particular year, and the one Bernard often recalled with immense nostalgia, a meeting had been arranged between the owners and the union representatives. The meeting was held in one of the plush hotels situated in Swindon. A neutral venue, well away from the hustle and bustle of the two incinerator facilities, and well away from the main offices in London, without the fear of frequent interruptions. Despite it being well before my time with Chemkiln, it is a meeting I am very well-acquainted with, for Bernard, especially after having consumed a few beers, loved relating the tale and repeating it in great detail. Sometimes, I even feel as if I had participated myself. Bernard's story, often fuelled by alcohol, proved to be his regular anecdotal, party piece. Obviously, an enjoyable and extremely memorable period in his life.

Before the meeting, Bernard and his union colleagues anticipated an argument concerning their forthcoming negotiations, predicting one almighty confrontation. As is the customary procedure with wage negotiations, the unions invariably aim higher than the actual pay increase they are looking for. Conversely, the management usually offer lower than the amount they are ultimately willing to pay. Such is the traditional format. The opposing factions discuss the amounts, invariably, both parties finally agree on a figure in the middle and somewhere near to where each party hoped to arrive in the first place. It is like a well-orchestrated, choreographed ritual. At the time of these negotiations, the average pay increase nationwide tended to be in the region of three to four percent. The union deputation initially intended requesting a six percent increase in salary across the board, anticipating an offer in the region of two percent from their employers. By the normal process, protocol and etiquette, they should arrive at approximately four percent, but only after following some stiff, intense arguments, bartering and discussions, both sides giving the appearance of holding firm yet exuding the pretence of willing to compromises lightly.

The union representatives entered the conference room which had been hired for the meeting. Upon entering the room, they discovered the management team already seated on one side of a huge mahogany conference table. The union members took their seats opposite. After the initial greetings and introductions, Alex Hedges, the MD, spoke first.

'We are going to make you an offer of a pay

increase for the next twelve months.' He said, while at the same time quietly acknowledging his business partner sitting immediate to his left, indicating they were both of one accord on this point. He continued,

'It is not open for discussion, and if you refuse, we will withdraw the offer, and there will be no increase for the next twelve months.' At this instant, he stopped to pour himself a glass of water. The union members looked at each other, girding their loins for the anticipated, full-blown, hostile confrontation. After taking a large sip of water, the MD continued with his monologue.

'So, because the company has had a reasonable year, both with turnover and profits, we are offering all your members, and non-union members alike, across the whole spectrum, a pay increase of eleven percent, with no strings attached.'

Alex remained silent while the union representatives looked at each other completely dumbfounded. They had all anticipated a much lower offer. This amount proved to be way above all their expectations, taking the wind completely out of their sails. Stunned, being insufficient an adjective describing their feelings or expressions. Most other employers nationwide tended to be offering around three percent, but it usually came with a caveat, and, some sort of proviso, such as increased productivity or reduced overtime.

The MD observing their confusion and incredulity, smiled to himself, realizing the disorientation he had caused. He also felt slightly pleased with himself.

'Would you like some time to discuss the offer amongst yourselves?' he then asked.

From a personal aspect, Bernard's mouth dropped, but he did ask for the reassurance there would be no caveat requiring extra productivity. A reassurance which Alex Hedges re-iterated without any equivocation or hesitation.

The shop stewards trooped out, vacating the meeting room. Following a quick discussion, they unanimously agreed to accept the outstanding, extremely generous offer. Returning to the conference room they informed the owners of their decision and agreement to accept the offer on the table.

'I thought you would,' Alex Hedges replied, his smile having now morphed into a broad grin, which exuded from his face, knowing full well the confusion he had generated amongst the union representatives. He then added, 'We have had an extremely good year and we want everyone to have a generous slice of the cake and share in our good fortune. Now let's get to the important business, shall we all go to the hotel bar and have a drink. There's a tab open and it's waiting to be broken into?'

Bernard often recounted this incident with a glazed look in his eyes, often extolling the virtues of his former, generous employers.

'Ah! They were such happy days and good times, with fantastic pay. Probably the best period of my working life.'

Alex Hedges and Gerald Smith proved not all employers are bad. They knew how to keep the work force happy and get the best out of them. In return, they received unflinching and dedicated loyalty. The two owners needed to have their employees fully on their side. They realized their staff had to contend with vast amounts of adverse publicity together with abuse and criticism from the local residents being continually hurled at them. Within a few years of acquiring the incinerators, their generous nature and business sense paid huge dividends for both partners. They sold the two incinerator operations to another multi-national waste disposal company for the grand sum estimated to be in the region of eighty million pounds. Just at the time when the PCB side of the business began a slow decline and the money generating bubble burst. Not a bad return on two million and selling up at an opportune time.

Bernard informed me Alex Hedges subsequently bought a large yacht, eventually circumnavigating the world on the back of his well-orchestrated business windfall.

During that time in its history, the company did come in for some fairly adverse publicity for the chemicals being disposed of at the facility. One of the most infamous toxic loads of chemicals involved a Russian ship called the 'Karin B.' A ship fully laden with a highly toxic pesticide called Kepone, with its origins in Canada and destined for the South Wales incineration facility. Of course, ultimately, the press and media became aware of the consignment. The media fuelling the demonstrations which became a frequent event outside the gates of the facility as well as the docks where the Karin B intended to off-load the shipment. So much opposition became generated by the consignment being shipped on the Russian vessel, the government ultimately caved into public pressure, finally refusing permission for the Karin B to dock in the United Kingdom. Other countries also refused to accept the waste. The Karin B seemed doomed to eternally roam the seven seas like the mythological Flying Dutchman.

I do not know what eventually became of the toxic consignment, although, I have my suspicions it somehow ultimately made it back to the UK for disposal by other nefarious, clandestine means.

CHAPTER 3

A fair number of the projects carried out by the General Services department on behalf of the company, necessitated a considerable amount of travelling, entailing being away from home for extensive periods of time. Home, for me by this stage in my life, and working career, meant living with Stella, my partner. Unfortunately, by this juncture in our relationship, things had somehow managed to turn extremely sour between the two of us. Stella worked most evenings and weekends as a part time bar-maid, appearing to relish the time she spent working behind the bar, as well as obviously savouring the time I spent away from home with my new job. My time away posed no problem concerning our disintegrating relationship. In fact, it helped, with both of us appreciating the enforced, prolonged periods of absence and being away from each other. I also used the time spent away from Stella to enhance my love life, as the reader will later discover.

Many anecdotes and stories circulated concerning the General Services department, a department I now found myself working for and a part of. Many of the fables, accounts and tales recounted within the company presented the department in an almost legendary, mystical, reverential way. Some of the stories imparted, verging on the unbelievable and the preposterous. However, I can personally testify, all true.

Before it became known as General Services, the department originally bore the title of Special Operations. I must admit, this latter name appeared more fascinating and alluring, conjuring up all sorts of adventurous, exotic, macho images for the small elite department. Images such as people being parachuted or flown by Lysander light aircraft into enemy occupied

France during World War II. Or, alternatively, visions of elite, special forces personnel, all dressed in black, their faces hidden by balaclavas or gas masks; abseiling at high speed down ropes, throwing smoke and percussion grenades into an embassy before propelling themselves through the shattered windows. Meanwhile, Kate Adie, with great verbal dexterity, describing the proceedings to millions of television viewers glued to their television screens, safely ensconced in the confines of their homes, utterly hypnotized and mesmerized by events unfolding before them on their small screens. Hence, in comparison, the new name, General Services, appeared; non-descript, ordinary and bland, proving to be not quite as evocative and exciting as the original departmental name, failing lamentably in generating the same amount of mysticism, kudos and gravitas. The company often required the department to take on difficult, dangerous chemical tasks and projects which others, within, or outside, the company demonstrated an unwillingness to tackle. The name Special Operations did seem a far more apt name for the function which the small elite department performed within the company.

I had only been working for the company a couple of months when this reputation which the department exuded became apparent. One, frosty, cold, miserable December weekend, I slipped on an icy pavement, cracking my right cheek bone. This accident resulted in me ending up with one hell of a shiner to my right eye as well as grazes to my knuckles. Going to work the following Monday, I felt embarrassed at my appearance, my face exhibiting one multi-coloured, immensely swollen, half-closed eye. Of course, I received a lot of ribbing concerning my supposed pugilistic skills. Everyone immediately jumping to the wrong conclusion, with comments such as, 'so what did the other guy look like after you finished with him?'

One of the accountants from the Finance department even remarked, 'Glad to see you are upholding the reputation of the Special Operation department,' still referring to the department by its former, original name. This last statement by the accountant epitomised the reputation which the small department exuded throughout the company.

This change in the departmental name came about mainly because of two incidents which occurred in the early nineteen nineties. Both incidents followed in the wake of the first Gulf War conflict.

The American forces maintained military bases in the south-eastern region of Turkey, their locations quite close to the Turkish border with Iraq. The bases possessed hangars and runways for military aircraft to over-fly Iraq, preventing the then incumbent dictator, Saddam Hussein, using his small, virtually ineffectual air force, keeping it grounded. As part of the infrastructure for setting up the bases, the Sea Bees, the name given to the American Engineers, utilized old transformers to provide the electrical system supplying power to the base. Being quite old, the transformers contained toxic PCB, (Polychlorinated Bi Phenyls), transformer oil prohibited for use from the late seventies onwards.

The Sea Bees wanted to replace them with more up to date, efficient transformers. The only way of destroying this chemical completely and safely necessitated the use of high temperature incineration. American environmental law possesses its own idiosyncratic regulations concerning the international movement of PCB. The Americans consider it quite acceptable for their armed forces to pollute and contaminate other countries with this chemical. However, once taken out of the United States, Federal Environmental Law prohibits the armed forces bringing the electrical transformers and the carcinogenic oil, back to the USA. This meant before the closure of any bases, the transformers had to be emptied, flushed with inert mineral oil, the PCB, together with the flushing's, then pumped into drums, for shipment to any country possessing high temperature incineration facilities, along with the empty carcasses of the transformers. The United Kingdom being one of the few countries capable of providing this service, through two private waste companies, Chemkiln and another company called Cleanaway.

The Sea Bees needed to remove some transformers from one of their bases and Chemkiln won the contract, sending a team from Special Operations, the name still used at that time, to perform this ecologically beneficial task of bulking the oil and arranging shipment back to the UK for disposal. Ensuring all transportation rules and regulations for the packaging of waste conformed to the international laws for transportation by sea, compliant with the IMDG Regulations. That is to say, correct packaging and labelling. In fact, it is a very demanding and stringent international law, instigated following the horrendous maritime disaster involving the Titanic in April 1912. The Internationally agreed laws bringing into force many rules such as having enough life boats on ships for all the passengers and crew, as well as policies and instructions for the segregation and packaging of hazardous chemicals.

I apologize, as yet again, I start to go off on a tangent. Back to the story of getting rid of the transformers and the PCB. At that time, the company put Damien Phipps, the same Damien Phipps who would later become my boss, in charge as Project Manager. Upon their arrival when passing through the Turkish customs, the team endured peculiar looks from the Turkish customs officials. The officials carefully studied the additional paperwork provided to speed the Special Operations team's progress into and through

Turkey. This additional paperwork indicated the team belonged to a department called 'Special Operations,' employed by a company called Chemkiln and working on behalf of the American forces. The only thing which highlighted and registered with the Turkish customs personnel tended to be the words *'Special Operations,' from the UK, working on behalf of the Americans.*

Coincidentally, the location of the Welsh incineration facility happened to be not too far from the location of the famous, or infamous SAS (Special Air Service), based on the outskirts of Hereford, also often referred to as Special Operations. Consequently, a thoroughly erroneous and incorrect scenario of assumptions began to slowly formulate and synthesize. *Special Operations from South West Britain, currently based on an American Air force base, located just a few miles from the Turkey-Iraq border.* Two and two were being incorrectly added together; coming up with five, six or even seven.

On their first day, while journeying to the transformers on the base, the team suddenly discovered themselves being used as target practice by fanatical, die-hard, Iraqi snipers. The snipers, shooting at the team whenever the opportunity presented itself, having somehow speciously been misinformed relating to the SAS being present on the base. Stories began circulating within Iraq about the SAS being on the base for the specific task of setting up a highly technically advanced rocket launching system. The Iraqis assumed the ancient electrical transformers to actually be those rocket launchers. Fortunately for the team, the transformers proved to be just out of range of the snipers' rifles. The journey to and from the work location posed the most danger. Henceforth, the team's daily commute to and from the worksite comprised of the team, attired in flak jackets, lying face down on the floor of an amour re-enforced Humvee vehicle. This safety ritual of self-preservation became the normal commute to and from the work site, persisting throughout the duration of the project.

At the same time, the departmental name brought about another incident which also helped bring about the name change. The team stayed on the base for the duration of the project. During the evenings, the Americans permitted the team to avail themselves of the ample facilities provided in the officer's mess. Consequently, the General Services team, (sorry, Special Operations) enjoyed themselves, taking full advantage of the facilities and hospitality being provided. On their first evening in the mess, the team comprising of Damien Phipps, and the operators; Bernard Evans, Steve Williams and Ioan Winston, walked up to the bar and ordered beers, lager or cider. Their preferences varied on this particular subject. The barman poured out the alcoholic beverages, then proffered them to the parched team. Damien, being the supervisor, and the one holding the expense account, presented his US dollars to the barman. The barman promptly declined the money being offered to him, quickly explaining in his laconic, southern American drawl, the drinks had already been paid for, before adding further information. It seemed he had received deliberate and precise instructions not to take any money from the team for any beverages, alcoholic or otherwise which the Special Operations team consumed during their time spent on the base. The team enquired who was paying for their drinks, so they could thank the individuals personally. The barman then pointed to a group of British RAF pilots sitting in the corner. The aviators were also stationed at the American base and part of the international peace keeping force.

After picking up their drinks, the team walked over to the pilots who appeared to be enjoying themselves, relaxing after over-flying Iraq some hours earlier.

Damien being the, team leader spoke to the pilots first.

'Hi guys, Sorry to interrupt. Thanks for the drinks, but why are you paying for us. You don't know us from Adam?'

The pilots immediately stopped chatting and looked up at the Chemkiln team. The team had the distinct impression they were being studied extremely closely. The Squadron Leader and head of the team of pilots replied first.

'You're the SAS, the chaps who are going to get us out of Iraq if we are careless enough to get ourselves shot down over there. We just want to keep in your good books, that's all.' He spoke these words with a disarming, personable smile in his precise, clipped BBC accent. All the Special Operations team looked at him totally bemused.

'You're totally mistaken,' replied Damien, 'We're employed by a hazardous waste disposal company, specialising in high-temperature incineration. A company called Chemkiln. We're here to dispose of some electrical transformers and the toxic PCB oil.'

The spokesman for the RAF pilots looked at him, whilst at the same time tapping his nose and giving every one of the Special Operations team a knowing wink.

'We realize you boys have a front to keep up, with your cover story to maintain. We'll hear no more about it. The drinks are still on us.' The discussion continued for a short while longer. Yet, despite the team's continued assertions and protestations, the RAF pilots resolutely refused to believe they did not actually belong to the elite SAS.

The team stayed on the base for almost two weeks. However, towards the end of the project; they decided to push out the boat as far as the drinking was concerned. Damien, never one to miss out on an opportunity for free drinks at someone else's expense, went overboard. On the last evening, he became totally inebriated, before finally collapsing in a drunken stupor. The team had to carry him to bed and gently tuck him in. Unfortunately, it did not end there, with Damien being unable to get up in the morning, suffering from one monumental hangover.

This resulted in the rest of the team having to cover for him. Unluckily, for Damien, news later percolated back to the Chemkiln offices concerning his inebriated state, and his inability to function the next day. I suspect this information came via the Americans. Because of his drunken, noisy behaviour in the officers' mess, Damien received a reprimand. The company had no outright proof of how bad Damien's drunken condition proved to be, and, if it was indeed the reason for his inability to work the following morning. The team, loyal to the end, covered for their, recently employed, rookie supervisor, substantiated his story that he had been unwell and not hung-over as the American base commander had informed the company.

After the team had been continually shot at by fanatical members of the Iraqi army and allegedly encouraged to become thoroughly intoxicated, courtesy of Her Majesty's Royal Air force, resulting in the Supervisor being unable to function properly. All of these events transpiring because of a departmental name. As a safety precaution for their workforce, an executive decision, to change the name from 'Special Operations,' to the less adventurous sounding, 'General Services,' became implemented. Although other departments and personnel within the Chemkiln operation, mainly, because of historical habit, still continually referred to the department as 'Special Operations.'

What's in a name? The above events show the possible consequences.

CHAPTER 4

Working for the department could be quite dangerous, especially while working abroad. When it came to encountering danger, Rupert St John Smythe usually tended to be involved somewhere along the line. Rupert had an extremely interesting and courageous personal story.

When I first came across Rupert during my job interview, he had been in remission from cancer for several years. During his terrible illness he had virtually been at death's door, requiring all the usual treatments; radiotherapy, extensive surgery, chemotherapy, the whole gamut of the cancer treatment regime. From the stories related by the other members of the team, Rupert had been riddled with that terrible, cruel, unforgiving disease. It looked doubtful as to whether he would ultimately survive his illness. He had cancer of the liver, lung, lymph glands and God knows where else in his body.

If nothing else, Rupert proved to be a fighter, coupled with the fact, the cancer cells in his body responded extremely well to the chemotherapy. The cancer cells regressed, eventually dying away after being subject to the brutal assault by the treatment. I recall Rupert and myself visiting Velindre hospital in Cardiff. Velindre being one of the main cancer treatment hospitals in the vicinity. The reason for our visit, to evaluate some of the outdated laboratory chemicals intended for disposal. While there, inspecting the chemicals, one of the consultants paid us a call, and evidently overjoyed at seeing Rupert, after being informed of his presence in the hospital. He kept telling me how Rupert had become his star patient, and a beacon of hope to other cancer sufferers in the hospital. The surgeon, at seeing his star patient, obviously in the picture of rude health, felt unable to contain his obvious enthusiasm. Rupert exhibited discernible embarrassment and unease at being the centre of all this unwanted attention from his consultant surgeon; humbly attempting to laugh it off. The esteem and reverence with which the consultant and other doctors held Rupert, was palpable. I felt amazed, Rupert rarely talked about his ordeal and suffering experienced during his illness, which the doctors at Velindre happily explained in great detail.

During the course of my time working for Chemkiln, Rupert finally decided to '*pop the question,*' and marry his long-time girlfriend Lucy. They made a great couple. Lucy displayed a great personality, exuding an immensely friendly disposition. During the wedding evening, Rupert and a few of his old friends intended performing 'The Full Monty' routine. Although, not going the whole hog, with the all-male participants only stripping down to their underpants or swimming trunks. Everyone in the department had been invited to the wedding. I also went along with Stella, and it was one of the rare occasions we actually went anywhere together. All of us were well aware about the huge scar on Rupert's front torso, a memento from the surgical operations he had to endure. Although, only a few people had actually seen it.

When it came time for the performance, everyone's eyes became riveted on Rupert's front and stomach. The whole wedding party could not extricate their eyes from Rupert's torso. When he eventually stripped off his shirt, the loud gasp which emanated from the audience became discernible. My mouth dropped as I looked at the huge scar on his stomach. Rupert's front had an inverted cross displayed on it, with the top of the cross just below his navel. The arms of the cross extended the whole width of his body. The base of the cross, paradoxically, at the top of his chest, terminated just blow his oesophagus. From a distance, the actual width of the scar also appeared to be quite large. I became transfixed at the evidence of the huge trauma which Rupert's body had been forced to endure, now displayed to everyone present in the room. From that day onwards, Rupert had my complete admiration for both his courage and his humility concerning his illness. As a postscript, Rupert and Lucy are still together all these years later.

Many years prior to me joining Chemkiln, the department sent Rupert to Somalia; working on behalf of the UN, in order to assess the costs for the disposal of miscellaneous pesticides, such as DDT, which had become banned in Europe and the US, the former major users of the pesticides. The manufacturers of the pesticides sold their excess stocks to these third world countries at lower prices in order to dump them. However, after a year or so, these pesticides eventually became banned from use worldwide. Third world countries, particularly those in Africa and Asia, now possessed huge quantities of the prohibited pesticide which could now no longer be used. Not only that, they needed to be disposed of, using high-temperature incineration, the only environmentally safe way to completely dispose of the toxic pesticides. This disposal cost was being funded by the United Nations Environment Programme and the World Health Organization. Although, the big chemical corporations had a moral obligation to foot the bill, an obligation they generally ignored or only paid lip service to. Rupert had almost finished his costing appraisal, and on the point of heading back to good old Blighty, when he received a phone call from his boss, Jim Beam. Jim instructed Rupert to go further into the interior of Somalia to evaluate some more pesticides for disposal.

At that time, the country was experiencing a bitter civil war, one of the many which spasmodically occurred throughout the decades in that barren country.

Unfortunately, as with all civil wars, it developed into a bloody, vicious, murderous affair, with no quarter asked for or given by either side. As the pesticides in question unfortunately transpired to be in the interior of the country, Rupert had been informed he would be provided with a UN escort to ensure his safety in getting to the village in question. Unfortunately, this information proved to be inaccurate, if not misleading.

Rupert had been instructed to meet a Captain from the UN ground forces stationed in the area. Early the following morning, they met in the foyer of his hotel, after which, the Captain accompanied Rupert, who had been provided with a 4 x 4 vehicle, to the outskirts of the city, where a platoon of blue helmeted UN troops waited in a convoy of vehicles. The convoy set off for the village where Rupert had been instructed to perform his assessment and evaluation. The village, regrettably, proved to be located a couple of hundred miles away, and situated in some of the most inhospitable terrain and climate in the world. Because of the lack of maintenance performed on the roads due to the internal conflagration, the journey invariably took hours. Finally, the convoy stopped at an isolated crossroads. The UN captain left the air-conditioned confines of his vehicle, alighted athletically to the ground and slowly approached Rupert, who had been travelling alone in his hired SUV. The Captain then spoke to his charge.

'The village you want is approximately fifty miles down that road,' he said, while at the same time pointing in the direction of one of the barren, dusty roads, leading into a bleak, haze shimmering, distant horizon, and heading away from the intersection into oblivion.

The officer continued. 'Unfortunately, despite being a UN peacekeeping force, we are considered fair game by all of the belligerents, and often subjected to pot shots being taken at us from all the factions involved in the civil war. The safety of my men is paramount, and I am unable to escort you any further. I suggest you consider this to be as far as you are able to travel safely, and recommend you return with us, back to the capital.'

Now most sensible people would have replied, 'Sounds a first-rate idea and like a good plan to me. I'm coming back with you to the relative safety of the city and civilization. Sod the pesticides!'

I emphasise the fact, most sensible people. However, Rupert St John Smythe did not fall into the realms of sensible people, and decided to carry on regardless, despite being on his own, and without the protection and company of the blue helmeted UN troops. The Captain, from all accounts, tried desperately to dissuade his charge from carrying out what he considered to be a foolhardy risk, but eventually gave up the discussion, returning, with his troops, back to the capital.

I must explain, although the reader has already probably deduced the fact. Rupert possessed none of the attributes or the appearance of your action heroes such as; Bruce Willis, Arnold Schwarzenegger, Steven Segal, or John Wayne, the Duke. So, his act appeared even more bizarre. Rupert St John Smythe looked nothing like any of these film stars and action heroes. As cited before, Rupert measured in about 6 ft.7 inches, very thin and lanky, with hardly any fat on his bones. In the vernacular of the day, your typical 'nine stone weakling.' He wore thick lensed spectacles which resembled the bases of coke bottles, surrounded by frames which were probably the cheapest the National Health supplied. The same type of glasses he probably wore years later at my interview. One could not envisage Rupert as being the sort of person, single-handedly, taking on members of an armed, vicious, blood thirsty, psychopathic guerrilla force.

Yet, despite all his physical shortcomings, Rupert, in his own inimitable way, carried on regardless, heading down the dusty, shimmering, barren road towards the remote village, surrounded by a rebel army. That was the last the Captain saw of him, with Rupert heading one way, and the UN forces heading in the completely opposite direction.

Now, try visualising the scene back home, and after a considerable period of time had passed. Rupert had been gone for about two weeks, without any form of contact from him. Rupert's family constantly phoned Chemkiln to discover the whereabouts of their beloved relative. At that time, Rupert was not married, so the people enquiring about his whereabouts tended to be his mother, father and siblings. All sorts of recriminations began flying about. How dare Chemkiln send their beloved son/brother to some God forsaken, war torn, barren, inhospitable part of the world, without any form of protection. In fairness, their arguments were well founded. We are talking about the early nineties, without many forms of the modern-day technology, which today, we now take for granted. With no satellite links, mobile phones or Skype. Maintaining contact with Rupert proving nigh on impossible.

Then suddenly, one morning, a few weeks later, completely unexpected and unannounced, Rupert turned up in the Special Operation's office. He appeared to be in the peak of health, wonderfully tanned and without a scratch. Much to the utter amazement of all those present in the small open plan office. He had not phoned from the Somalian capital, just in case the department instructed him to go to another far-flung region of the country, before allowing his return home. So, without contacting the office, he caught the first available flight back.

Rupert related the stories of his adventures. Apparently, on his way to the village, Rupert had come across one of the armies involved in the civil war. The local War Lord, or perhaps a more apt description being, local bandit leader, took an immediate liking to the eccentric Rupert, who, it must be said, is very personable, proving difficult not to like. The local leader led Rupert to his village, providing him with shelter, food and other amenities. Rupert has never been forthcoming about what other 'amenities' the warlord supplied him with. Every time I posed the question, he always exuded a wry smile, denying anything inappropriate. He seems to have had quite an interesting time, despite the hardships, which is all that can be said. His adventure ended successfully, and without any mishaps, which is the main consideration.

The offending toxic pesticides were never removed. Well, not by Chemkiln anyway.

Rupert had quite an eventful and interesting career whilst at Chemkiln. If anything happened, you could be certain he tended to be involved somewhere along the way.

Chemkiln received a phone call from one of the local waste disposal authorities in Weybridge Surrey. A very '*well to do*' area of the Home Counties. It all began when a local private landlord had not received rent from a tenant living in one of his rather plush houses. The rent had been outstanding for a considerable period of time, and the landlord thought he had better pay a visit to collect all the monies due to him. He went to the house, and after knocking the door for some time, using his landlord's key, decided to go in. Upon opening the front door, some twenty-five litre drums, containing solvents and paints fell onto him. The house had been used as a disposal outlet by the tenant, who had provided his landlord with a false name and address, in order to cover his tracks. The tenant had filled the house from back to front with all types of waste; paints, acids, solvents, alkalis. He charged people for their removal, using the house as the disposal site and dumping ground. He had even managed to fill the garage, not leaving one square foot of unfilled space, before doing '*a runner.*' The tenant had not even paid rent for the time he was there. This was one sharp, devious, unscrupulous operator. In those days disposal of waste was not highly regulated and monitored as closely as it is today. The landlord was at his wits end. He contacted the local waste disposal authority; who, in turn, contacted Chemkiln to dispose of the chemicals in a proper, authorized and legal manner.

The company decided to send Rupert to the house, to carry out an inventory of the waste before shipment by road, then disposal at the South Wales facility. To carry out a proper, thorough, inventory, Rupert pulled each drum out of the house and commenced putting them onto the long driveway to log and perform the comprehensive register. About an hour or so into the task, all hell suddenly broke out, with the wailing of police cars and fire engine sirens disturbing the serene tranquillity of the salubrious, exclusive suburb. At the top of the driveway, the police cars screeched to a halt, and a few policemen jumped out, immediately hurtling towards the startled Rupert.

Before he could utter a word, they put him up against the wall of the house, instructing him to put his hands on his head, then pulling his legs wide apart. After which, they then heavily frisked him, cuffed his hands behind his back, then frog-marching him down the driveway, before manhandling him unceremoniously into the back of a waiting police car. The Firemen immediately began closely inspecting the drums on the driveway. While they were doing that, the police car, containing the bemused Rupert, shot off at high speed, heading towards the local police station, where they subjected the unfortunate chemist to extensive, aggressive cross questioning. Following a frantic phone call to Chemkiln, as well as calls being made to the local waste disposal authority, the security authorities eventually released their prisoner.

Rupert's temporary incarceration came about due mainly to the residents in the locality. These were people who lived a secluded, buffered, privileged life style, in the '*well to do*' area. Predominantly high-flying, Bankers, Stockbrokers and Barristers. One can only imagine their incredulity at the sight of a tall, lanky individual, wearing white, disposable Tyvek overalls, a gasmask, rigger boots and green nitrile gauntlets suddenly depositing numerous drums of miscellaneous chemicals on the driveway of an extremely salubrious house, in a very exclusive area of Weybridge. Perhaps the person in the Tyvek overalls was a terrorist sorting out his explosives? The curtain twitchers first port of call was the local cop shop, asking them to come and arrest this undesirable person who was polluting their exclusive area with toxic chemicals, or perhaps even preparing to blow up the neighbourhood. Completely unaware Rupert was only trying to expedite their removal from the area.

I believe Rupert had to make an appearance in court when the authorities eventually apprehended the perpetrator of all the trouble and disruption, with Rupert giving evidence. Wherever there was trouble you could be sure to find Rupert.

Besides Turkey and Somalia, there were other dangerous parts of the world in which the members of the General Services / Special Operations department sometimes worked. One of these was Yemen, another inhospitable country in the world in which to work. There were very few amenities; the country had not really embraced the twentieth century, with outdated facilities and equipment. Once again, there were prohibited pesticides in the country which required removal and disposal. Once again, Chemkiln had been called in by the UN to perform the task. The project proved to be quite extensive, with a few sites throughout the country. The project was estimated to last a minimum of two months.

After discussions with the members of the General Services department, it was agreed the project should last no longer than six to eight weeks. More than enough time to spend in that dry, arid, inhospitable part of the world. The political atmosphere also tended to be very unstable, which did not help the situation either. It was agreed, within the department, the job would be done in two shifts. The first shift would stay for about three weeks. They would then be relieved by the remaining members of the department. The first shift would then return home for a well-earned rest.

Upon their arrival, the team discovered conditions were much worse than the section had thought, with limited shelter and water supplies at the areas where the pesticides were located, miles from anywhere. The teams had to wear protective impervious overalls and masks with heavy duty boots and nitrile gauntlets. Not ideal clothing with temperatures well into the nineties, or low hundreds Fahrenheit.

The team would take a break during the weekend. Apparently, the taxi drivers who occasionally took the team to the location, frequently stopped on the way to pick foliage, which grew in shrub trees on the side of the road. The foliage, called Khat (Qat), pronounced cat, Yemen's cheap, readily available, legal equivalent to cannabis. The drivers would often stop, collect some leaves, then chew them whilst driving, slowly becoming totally stoned and spaced out. Not an ideal state to be in while driving around the dangerous, narrow, mountainous roads, often leaving the members of the Chemkiln team slightly traumatised. As alcohol was banned, with Yemen being a strict Muslim country, Khat (Qat) appeared to be the only way the indigenous population could indulge in some form of escapism from the reality of their poor wretched lives. Sometimes, after the drivers had chewed too much Qat, they would park their taxi and fall asleep. All these unexpected stoppages helped prolong the work.

The work progressed slowly, with toxic pesticides being bagged, or pumped into the approved containers for shipment by sea, ensuring the IMDG Regulations were strictly adhered to for the transportation of the toxic chemicals back to the UK.

Much to the relief of the first team, the first three weeks stint came to an end. They quickly packed their bags, met the support team, informing them of the work done and the work left to be done.

They then promptly left on the first regular flight back to Europe.

Now Rupert, who else? He had been the first Team Leader, and left all the hotel bills to be paid for by Liam, the new Team Leader, and who also had an expense account. Liam signed into the hotel, only to be met by a disgruntled Hotel Manager, who was not happy the first set of guests at his hotel had left without settling the bill, demanding to be paid at once, and in cash. Liam protested he did not have enough spare cash, informing him he would settle the bill with his credit card. The Manager became insistent, he wanted cash. Nothing else would do. Despite Liam's protestations, that he did not have enough cash to pay for all the hotel rooms for the whole six weeks. Within a few minutes, armed soldiers entered the hotel and put all the second team of the department under arrest, and at gunpoint. Liam had to make some frantic phone calls to the Chemkiln's office, instructing them to send a banker's draft through to the hotel, to pay the outstanding hotel bill, and as quickly as possible. He also asked for additional money to be sent to pay the bill in advance for the next few weeks.

The whole experience completely unnerved Liam. Upon his return to the UK, he asked to be transferred to another department within the facility; no longer wishing to work abroad for the department.

There are other parts of the world not so dangerous. I was once instructed to fly to Jamaica and evaluate some pesticides for disposal. I travelled to Heathrow to board my British Airways flight to Kingston. Upon getting to the departure desk, the girl informed me to phone the office, as it appeared the job had been cancelled. I asked the young employee whether she could say I had boarded the aircraft before being given the information. I had never been to Jamaica and wouldn't have minded a quick trip to the exotic Caribbean island. While asking for her help, I gave her a knowing wink. The young girl smiled back at me, while at the same time shrugging her slender shoulders.

'I would dearly like to help you, but unfortunately, your company has already cancelled the ticket.'

To this day, I still haven't been to Jamaica. Ah well.

CHAPTER 5

Albert could be one of those irascible, annoying, bigoted individuals, continually nosing and interfering in other people's business, with an inclination to snitch or grass on any of his colleagues concerning any mistakes or misdemeanours on their part. Not too dissimilar in that respect to Cedric Hughes, the Maintenance Engineer who worked for my previous employer Hyperwaste. When I first encountered Albert, he appeared to be gently cruising towards his retirement, just a few years down the line. Consequently, the company reluctantly tolerated all his idiosyncrasies and foibles, waiting for him to eventually retire.

A couple of years earlier, before I joined Chemkiln, it seemed Albert possessed a reasonably affable, easy going temperament. However, an incident occurred which affected and upset him. The other guys in the department somewhat cynically suggested the event somehow twisted and distorted his original, likeable persona. My opinion tended to be, the natural ageing process simply kicked in; Albert had basically turned into your typical, curmudgeonly, grumpy, old man. Then perhaps I am being the cynical one.

The episode which supposedly caused the transformation in Albert's personality, came about after the company had been requested to send some personnel to Pembroke in West Wales, where a council tenant had died quite suddenly and unexpectedly from natural causes. Chemkiln had been requested to inspect and evaluate the chemicals contained in the house. The dead man had no living close relatives; his nearest living next of kin being a brother-in-law and the husband of the deceased man's now dead, elder sister. It seems, the deceased man absolutely idolised and worshipped his sibling.

The recently departed man had one, all consuming passion; taxidermy, the stuffing and preserving of dead animals. Unfortunately, because of this unusual pastime, the council house contained a huge assortment of chemicals normally associated with this particular hobby. Materials such as; Formaldehyde (formalin), clays, solvents, resins, paints, tools, wool, plaster of Paris, soaps, detergents, and quite a few other highly toxic materials such as arsenic.

As instructed, Albert and Rupert drove down to the house, and met with the brother-in-law of the deceased tenant, who was accompanied by one of the council employees from the local Housing Department. Albert and Rupert were to carry out an initial assessment for removing and disposal of the hazardous chemicals.

Gingerly, the four of them entered the council house, only to be enveloped by the overpowering stench of dead animals, and the sickly odour of preservatives associated with taxidermy. After separating in the main hallway, they began inspecting the now deserted premises. The council wanted all the chemicals and stuffed animals removed from the property in an environmentally friendly, proper and legal manner, hoping Chemkiln would perform the exercise as quickly as possible. After which, the Housing Department fully intended re-housing some of the people already registered on their extensive housing list.

Albert decided to inspect upstairs, slowly ascending the steep stairway to the first floor. While he explored upstairs, the three others remained on the lower floor, looking through the numerous cupboards, most of which contained an impressive array of diverse preserved creatures. Following a brief interlude, the three men downstairs heard a blood-curdling scream; evidently generated by Albert. His piercing cry of anguish appearing to come from one of the upstairs bedrooms. Rupert, accompanied by the deceased's brother-in-law and the council employee, hastily bolted upstairs to investigate the reason for the horrific, spine-chilling scream so recently emitted by Albert. Upon entering the bedroom, from which the cry appeared to emanate, Rupert encountered his work colleague sitting on a large bed, situated in the centre of the room. Albert shook violently, exhibiting a look of deranged terror on his face. Rupert tentatively entered the room, after first looking all around before doing so.

'What's the matter Albert?' he enquired, in his slow, anally retentive, refined, English accent.

There, immediately in front of Albert, stood a tall, metal cabinet, its two large doors ajar, exposing the contents inside. Albert, with the index finger of his right hand, pointed towards the oblong metallic piece of furniture his finger and hand shaking uncontrollably. He managed to stammer two words with extreme difficulty. Whatever had traumatized Albert, prevented him from uttering any more words. Take it from me, anything which prevented Albert from his normal verbosity and oral diarrhoea, had to be out of the ordinary, and thoroughly dreadful.

'Llllook ttthere!' he stammered.

Rupert looked directly to where Albert's trembling finger pointed. To his horror, Rupert discovered himself gazing straight into the hazel brown eyes of a decapitated woman's head. A head which appeared to be staring intently and directly back at him. Rupert also uttered an unrestrained gasp of horror and shock at the sight which he beheld; instantly falling backwards, as if poleaxed by a large baseball bat. Without realizing it, he too ended up sitting on the bed next to Albert. The two men stared, open mouthed at the well-preserved woman's head. Both in total disbelief, each dumbfounded, and at a complete loss for words.

Almost instantaneously, the two other men entered

the room, looking immediately to where Rupert and Albert remained gawping, open-mouthed and speechless. The brother-in-law of the deceased man spoke first, after looking in the direction where the two men gazed, the two of them, mesmerised, like rabbits caught in the headlights of an oncoming car.

'Bloody hell! That's my wife. What the hell is her head doing there, she was supposed to have been buried eighteen months ago?' exclaimed the dead man's brother-in-law.

From that instant, total confusion reigned, with frantic phone calls being made to various regulatory authorities and law enforcement offices, to discover some semblance of reasoning as to why the head of a woman who was supposed to have been buried, eighteen months earlier, suddenly manifests itself in the house of her recently deceased brother.

Further inspection of the house disclosed other well-preserved parts of her body. Her torso, legs and arms had been secreted away in the substantial, voluminous drawers of large cabinets. Unfortunately for him, Albert also discovered the preserved left leg of the woman, which did nothing to alleviate his already fragile disposition. The four individuals had managed to stumble upon Pembroke's equivalent to Norman Bates.

Despite extensive enquiries, the authorities came no nearer to discovering the how's, why's and wherefores of the woman's body turning up in the council house.

Chemkiln reluctantly obtained the contract for removal and disposal of all the chemicals in the house, required for the bizarre hobby of taxidermy. All the chemicals were disposed of within a few weeks. Albert refused to have any more involvement with the project. He never set foot in the house again.

The numerous parts of the woman's body were collected and removed to the nearest morgue. A political tussle then ensued between the two Coroners departments of Pembroke and Torfaen. The suggestion had been proposed to dispose of the poor woman's body by incineration at the Chemkiln facility. The Coroner's department, based in Pembroke, argued, that because the body would ultimately be disposed of in Torfaen, then the authorities in Torfaen should deal with all the confused, intricate paperwork concerning the woman's death and her ultimate disposal. Whereas, the Coroner's department, based in Torfaen, conversely insisted, because the body originated in Pembroke, it should be the Pembroke coroner who should have the dubious pleasure of dealing with the situation. After much to and froing between the two Coroners' departments, The Pembroke Coroner eventually acquiesced to pressure, reluctantly agreeing to handle the case.

The chemicals, along with the stuffed animals, were disposed of fairly quickly, within a couple of weeks, as a matter of fact, incurring no problems whatsoever. I understand, the woman's body eventually underwent a post mortem, ensuring she had not died in any suspicious circumstances. Nothing untoward concerning the woman's death manifested itself. After much protestation by the husband, concerning the interminable delays, the woman was eventually reburied in the original grave where her body had been previously thought to have been assigned, and erroneously believed to have been incarcerated, almost eighteen months earlier.

Albert reminded me, on so many levels, of Cedric, my former colleague from Hyperwaste. One of those levels, his total self-belief he was completely irresistible to women, despite the fact he was in his sixties, bald as an egg, apart from the tufts of hair on the both sides of his head, and the owner of an extensive beer pot. He was also the possessor of ill-fitting dentures, which gave him the appearance of Red

Rum when smiling for the cameras. Both top and bottom sets, continually lolled, and travelled around in his mouth, particularly after he had applied insufficient Fixodent, to keep them in-situ. As if all this was not enough, when nervous, Albert incessantly sucked on those same dentures as if they were Werther's original toffees.

Yet, despite all of the aforementioned encumbrances and impediments, Albert held the inalienable belief, in his own mind, he was an Adonis, lusted after and desired by the fairer sex, who would do virtually anything to spend a night of passion with him. A legend in his own mind.

A few of us, including Albert, had been involved in a project decontaminating a laboratory for a major chemical company based in Tyneside. The project was nearing completion, and one evening we decided to have a celebratory blow out in one of the local pubs. There were a couple of middle-aged, attractive women sitting at the bar. Albert sidled up next to them on the pretext of ordering a round of drinks, but his main intention being to chat to them, if his luck was in, '*cop off*' with one of them, or, if his luck was really in, then possibly ending up with the two of them. He tried all his standard chat up lines. I must confess, he did seem to be ingratiating himself with the two ladies, who did seem to appreciate his humour. Although, it could have been purely out of politeness on their part. This chatting and flirting had been underway for some minutes, when two men, sharply dressed in expensive mohair suits, obviously, high-powered salesmen, suddenly plonked themselves down on the bar stools next to the woman and furthest away from Albert. Much to Albert's annoyance, they began slowly interloping into the conversation. One of the men then moved closer, and surreptitiously began slotting himself between Albert and the attractive woman sitting immediately next to him. The salesman began talking about the exquisite red Porsche which he owned, and how much he loved and cherished his beautiful, gleaming red car. The Porsche, so it appeared, proving to be his pride and joy. This was all too much for Albert, even to his own self-aggrandisement, he could see he was being outgunned, outclassed and outmanoeuvred. Albert did not own a Porsche. Apart from the new Rover 75 company car, his family car was a Ford Mondeo, not a Carrera. Besides, the two men were probably two decades younger than Albert, and approximately the same age as the women sitting at the bar. Yet another factor in the two salesmen's favour. No comparison; game over. Slowly, Albert excused himself from the conversation, picked up the drinks the barmen had put in front of him, and returned back to our table, where we had all been listening to the conversation, observing intently, while continually sniggering to ourselves.

'The cocky bastard owns a Porsche, I stood no chance,' lamented Albert, consoling himself with some alcohol, by taking an extremely large sip of his beer.

'We heard,' interjected Ioan, desperately trying to hide the huge smirk on his face, by gazing intently into his partly consumed pint of lager.

After Albert sat down, we all continued listening to the conversation taking place at the bar, completely enthralled by the mating ritual taking place between the two salesmen and the two ladies.

'May I see your Porsche?' Enquired the attractive woman, who had been previously sitting next to Albert.

Result for the Lothario, the salesman had been completely successful in his endeavour. It was sweet surrender, and he made it all seem so effortless. All this, adding to Albert's increasing annoyance.

'Sure, if you like, 'the salesman replied, smiling.

With that, he put his hand in his jacket pocket, then pulled out a miniature model of a red Carrera, which he then propelled along the bar in front of the two women. 'Fantastic isn't it. Isn't she a beauty?' he quipped rhetorically. Both woman immediately burst out laughing, evidently considering the salesman's subterfuge and patter highly amusing.

Albert uttered quite a few oaths, gulped down his beer, then retired gloomily to his room. It was just all too much for him. Outmanoeuvred by a smooth talking, smarmy, clever bastard. The two ladies and both salesmen left shortly afterwards. One can only surmise the eventual outcome.

Albert remained the subject of many anecdotes within the company, until he retired a couple of years later after this particular incident. Poor, deluded Albert.

CHAPTER 6

*O*ne of the many good things about working for Chemkiln, and being part of the General Services team, tended to be the amount of training courses I attended. Obtaining experience and instruction in an eclectic mixture of subjects such as; First Aid, Health and Safety, Breathing Apparatus, ADR for transportation of hazardous materials, IMDG, Miscellaneous Computer Software courses, team Building, Project management, Fork Truck Driving, Defensive driving, Report Writing, Handling Difficult People. The list goes on and on. In fact, having attended so many courses, I am unable to recall every single one of them with any form of total clarity.

Most of the man-management courses tended to be run by the Industrial Society, a renowned and well-respected consultancy company, known throughout the UK. When they first came to the plant the memo circulated indicated there would be two instructors lecturing on the course. I noticed one of the names as being Carl Davies and wondered if it could possibly be the same Carl Davies I once worked with, many moons previously, in the long distant past, during one of my earlier incarnations.

For I had once worked with a certain Carl Davies during my time as a Process Engineer, whilst employed by a Hi-Tech company called Repeat Controls. We had been informed the two instructors from the Industrial Society would be walking around the site, introducing themselves to people. Chemkiln fully intended enrolling as many of their staff on the course as possible. Most of their employees would all be attending a team building course at one time or another. The company felt that people like Damien, our department head, and some other individuals, certainly needed the benefit of the course. The only team

Damien belonged to was himself, being in the job for the greater glory of Damien. Team spirit did not manifest itself readily in Damien's particular world.

I could not wait for the two instructors to come into the office to discover if it was indeed the same Carl Davies I once worked alongside, during the former decade of the eighties. The same Carl Davies, who, together, with myself, helped resolve a problematic process issue, ultimately saving Repeat Controls, our employer, hundreds of thousands of pounds, or rather dollars; it being an American Corporation. Our joint action, and co-operation, helping us both acquire a large amount of brownie points with higher management.

Eventually, one afternoon, the course instructors finally came around to introduce themselves. Much to my gratification, the one instructor, indeed, turned out to be the same Carl Davies from all those years ago. He appeared to be slightly broader around the girth than when we last worked together. Indeed, as was I. His hair had become much greyer since our younger halcyon days, during those days of the eighties. The joys of getting old with the unrelenting ageing process. For, by this stage in our lives, we had reached the year 1996. Both of us quite a few years older from when we last worked together.

Carl recognized me immediately, despite my hair starting to wave, goodbye to my head, that is. We spent some time together, happily recalling and reminiscing about the good old days, talking about all the people we both knew, and those we remained in contact with. The memories came flooding back. Repeat Controls still operated, but nowhere near its size during the eighties, with barely eighty people left at the old site, compared with almost fifteen hundred when we both worked there. A sad loss to the community. It felt good seeing him again, we both brought each other up to date with our personal circumstances. Carl was now divorced, living on his own. I told him about Stella and myself living together, which shocked him somewhat, as he also knew her, but not of our relationship. Carl left Repeat Controls well before I did. During his last weeks with the company, he had been sent to the States for almost a month, on a fact-finding mission. Upon his return, he promptly handed in his notice. You had to admire his style, he had a month's trip to the States, paid for by the company, and then quits.

Finally, both he and Dave, the other tutor, left the office. Carl informed me, he would be at the team building course the following week.

The Team Building courses went very well, although both Carl and Dave Woods, the other course tutor, appeared unimpressed by some of the staff, particularly Damien, and a certain Jonathan North, who had once worked for BP. Jonathan tended to be a throwback to the days of the Raj, possessing a thoroughly racist and bigoted mentality, often disparagingly referring to Africa as Bongo Bongo land, talking about the natives having bones through their noses. Both Carl and Dave, as a sort of repost, in an intentionally and discourteous way, often referred to Jonathan as '*the Great White Hunter*.' Dave would visibly cringe every time he saw Jonathan, or unfortunately found himself sitting next to him during meal times. The company later issued a directive, informing all staff, 'anyone found making racist remarks within the company would be instantly dismissed.' I suspect Jonathan had a severe talking to before the directive was issued, warning him about his racist, derogatory remarks, no doubt, with both Carl and Dave, having a hand in it. Good for the two of them, I say.

I also attended other courses, such as, First Aid in Industry, run by the Red Cross in Newport. Everyone in the department attended a First Aid course at one time or other, it was necessary we did so. Working normally in teams of two, or more, and far from others, it was imperative everyone knew what to do, in case of an accident or an emergency. I am a bit squeamish, particularly when it comes to organs and various exposed parts of the human body. Despite this queasiness on my part, I must admit to finding the course thoroughly interesting and entertaining, due mostly, to Herbert, the senior course lecturer, who turned out to be quite a character. He would frequently simulate an incident or accident, performing, what can only be described, in an inarticulate way, as a sort of rubbery doll act, going through the motions of impersonating a person, for no apparent reason, collapsing limply to the floor. Whilst lying on the floor, Herbert would then casually pick his head up and enquire, what we would do next. I found the whole performance quite humorous, often quietly chuckling to myself.

Herbert showed us how definitely not to do head bandages. One in particular, where the result meant the injured victim resembled Bugs Bunny, with his large floppy ears, the bandages acting as the floppy ears. But, what I enjoyed more than anything, were the seemingly endless quantity of tales and stories about accidents he had personally been involved with throughout his first aid career. Howard would have the class, and the reader must excuse the obvious intended pun here, in stitches. His genial, good-natured disposition, enhancing the recounting of all these tales.

Our lecturer also explained, in detail, what to do with broken legs and the technique, which had been amended concerning not putting a solid splint, but to tie the break with foam or other more similar material. He related the story of a man whose leg had obviously been broken. The first aid team put a piece of wood as a splint on his leg. The wooden splint extended an inch or so longer than the victim's foot. He was rushed to the hospital by his colleagues. They lay him on a stretcher, pushing him along the hospital corridors, at a fair rate of knots. Now, one of those attending to the victim had been to that hospital a few weeks before, and knew the swing door swung inwards, or at least, thought he knew, the doors would swing open. However, he did not know. The direction of opening the doors had been altered, with the doors now swinging outwards, not inwards, having been reversed within a week or so, after his last visit there. As the team came to the first door at a high rate of knots, the wooden splint pushed against the swing doors, bringing both patient and stretcher trolley abruptly to a halt. With the result the splint ended up being firmly pushed along the inside of the patient's leg. What started off as a broken leg, ended up with something more serious for the poor victim, leaving him in even more pain than before.

Herbert had a plethora and seemingly limitless reserve of these first aid anecdotes concerning accidents he had been involved with during his time as a member of the Red Cross. I asked one of the nurses assisting on the course, if all the incidents Herbert talked about had really taken place. To which she replied, seemingly, rather morosely, and somewhat despondently,

'Oh yes! Herbert is like a magnet to accidents and incidents. They always seem to occur whenever he is about. We all dread going to any sort of event with him, whether it's a carnival, fête, football or rugby match, you can guarantee it will be a very busy day indeed!' I considered the nurse to be exaggerating somewhat, and most probably, as Del Boy would say, '*pulling my plonker.*' That belief quickly became dispelled the next day. The class normally went to a nearby local pub to obtain lunch. Before following the rest of the class, I phoned the work's office to check in, and get any updates on my work schedule for the coming week. After finishing the call, I walked along the one side of the street, attempting to catch up with the other pupils. There, on the other side of the road, I observed Herbert slowly walking along, obviously heading towards the small local bakery to buy some sandwiches or rolls for his lunch. I also observed a man walking on the same side of the road as Herbert, but going in the opposite direction, heading directly towards the instructor.

Suddenly, without any reason, and as if on cue, like an actor receiving explicit direction from Steven Spielberg, the man inexplicably collapsed immediately in front of him. Herbert, the ever consummate, professional First Aider, took care of the situation and revived the man, who had fainted for some mysterious reason. Well, fortunately for the man, he had somehow managed to collapse in front of the right person. After some first aid and attention, Herbert managed to fix the man and send him on his merry way, but only after instructing the man to make an appointment with his GP for a medical check-up as soon as possible. It did seem incidents occurred whenever Herbert tended to be about.

I am embarrassed to admit, I too fainted in front of Herbert. The incident occurred during one of his lectures. Having previously mentioned, I have a tendency for being squeamish when it comes to amputations and parts of the human body, organs and such-like, particularly after being removed from their correct position on the body. I proved this without doubt when Herbert began explaining what to do with a severed hand. He was in the process of carefully handling a plastic model of a dismembered hand, extremely realistically painted, complete with the image of dripping congealing blood and veins. Extremely too realistic for my liking. Herbert began explaining, the dismembered hand should be first put into a plastic bag full of ice, then transported to the nearest hospital for possible reconnection with its original owner. By now, my imagination had unfortunately started to run riot, overdrive, and virtually out of control. Almost immediately, I became hot, clammy and sweaty. The perspiration began running down my ruddy forehead and flushed face. I could also feel my shirt beginning to stick to my back, due to the perspiration, slowly trickling down my spine. Abruptly and unexpectedly, a dark curtain suddenly descended before my eyes. Immediately, the room went dark. I awoke, only to discover myself on the floor, surrounded by the class of my fellow students peering down at me, most with bemused expressions on their faces.

Some, however, I believe were desperately attempting to stifle their sniggers. Taking charge, Herbert stood directly over me, explaining to all those around him what to do next. I had suddenly, and involuntarily, become the practical lesson for that morning. Herbert kept instructing me to take deep breaths, while at the same time, explaining in detail what he was doing, for the benefit of the enthralled, slightly amused class. I was totally embarrassed, a grown man in his forties fainting at the sight of a false, red painted plastic hand. Thank God nobody from Chemkiln attended the same course, as I would certainly never have lived that one down. I would have undoubtedly been on the end of unremitting ribbing for some considerable time. The only saving grace being, I knew none of the others in the classroom, and unlikely to ever see any of them again. I continued the course, completely embarrassed, with Herbert looking periodically in my direction, frequently enquiring if I was okay. All of the class would then look at me, after he posed the question. I am certain most of the blokes were all still sniggering.

My concern, after eventually becoming qualified as a First Aider, was that I would not be able to perform and acquit myself, failing to attain the high standards expected of the Red Cross. However, I did have an opportunity to acquit and prove myself of being a worthy first aider, sometime later. Most of the General Services team were involved in the complete gutting, decontamination and disposal of laboratory equipment, working on behalf of a large Pharmaceutical company, based in Sandwich Kent. Staff working inside one of the research laboratories, developed symptoms of what is known as 'Chlor Acne.' With large sores, pustules and acne breaking out on their faces.

The laboratory staff were sent home and all put on the sick. The laboratory was immediately shut down, sealed off, quarantined and a major clean-up and decontamination project then set in motion. The pharmaceutical company wanted complete decontamination of the laboratory, with everything inside the laboratory destroyed by means of high temperature incineration, right down to the laboratory benches and sinks, having no idea what had caused the acne. We are talking a large amount of money here. 'Mega bucks,' as they say.

All the laboratory equipment, including expensive items such as GC's, XRF, Infra-Red, atomic absorption, fume cabinets, top of the range computers, printer's scanners, photocopiers, phones, faxes etc. In all, quite an extensive array of equipment, all destined for the Chemkiln incinerator. Everything had to be decontaminated with anti-bacterial agents prior to being put into two heavy duty plastic bags. The job made no easier by the laboratory being situated on the fourth floor of a tower block. The only access, made by cutting a very large hole in the outer wall, and a cargo container being put onto a platform, then used as a decontamination changing room, leading to the access hole. Attached to the platform, a lift for taking equipment up, and removal of all the contaminated laboratory equipment. Anyone entering the laboratory had to wear self-contained breathing apparatus, inside a full chemical suit. None of the team could come into contact with the contaminated materials in the laboratory.

I explain all this to show the extent of work involved in the project. The project requiring as many people as possible, taking on extra contractors who had worked for Chemkiln on previous occasions. I worked on the project for a couple of weeks, my services later being required elsewhere. During the duration of the project, we all stayed at the Jarvis Marina Hotel in Ramsgate.

I shall recount the one specific evening in question, which immediately springs to mind. The whole General Services team sat at a long table. Our meals had been ordered and we all indulged in convivial chat and banter. From the table, through a large glass partition, we were able to observe the hotel reception. I sat at the end of the table, one of the furthest from the glass partition. I looked forward to my evening meal, deeply engrossed in conversation. Suddenly, Albert, who sat to my right, got up, and suddenly began prodding me heavily in the shoulder,

'Come on Vinson!' he instructed brusquely, even though he was my peer in the department.

'There's a guy out there who looks as if he's in distress and about to faint. We're trained in First Aid, they may need and welcome our help and advice!'

I protested, arguing the point, that most of the hotel staff was probably also trained in First Aid as part of the health and safety requirements for the legal operation of the hotel. Bernard Evans, who sat further along, next to Albert, also on his right, likewise got up.

'Come on Vinson, he's right,' he added, in support of Albert,

'As Albert said, we can give them some help and advice.' I relented having a quick look at the guy in reception. I must say, he did look extremely distressed. The hotel staff and residents milled around him, which could not have helped his distress. The three of us went out to the reception area. Julie, the young and obviously panic-stricken receptionist, tried desperately to calm the patient. Unfortunately for me, out of our small group, I reached her first.

'Are you first aid trained?' I enquired.

'I attended my first course two weeks ago,' she replied rather gloomily. Before including as an addendum,

'This is my first incident.'

'Well there always has to be a first time!' I replied, smiling and trying to reassure her, almost as much as the afflicted resident, who, from his broken English accent appeared to be foreign, possibly German, and evidently very distressed.

'I can't breathe!' He kept protesting, in his guttural German accent, sweat, oozing from his pores; his breathing fast and extremely laboured.

'Undo your collar,' I instructed, which he did obediently and without any argument. I then assisted him in loosening his obviously expensive silk tie and the top buttons of his shirt.

Bernard added, 'Get him to sit down with his knees up.' So, I told him to sit in front of the reception desk on the floor with his knees up. Julie informed me an ambulance had been sent for and was hopefully, on its way. The anxious resident sat down, his breathing becoming even faster. Suddenly, his eyes rolled in his head and he abruptly keeled over.

'Put him into the recovery position, there's nothing we can do until he recovers, or the ambulance arrives.'

I said. We rolled him over into the standard recovery position with his head back and his airway open and preventing him from choking in case he suddenly started vomiting. I started feeling his pulse and noting his breathing. Now all this time, the two instigators of my press-ganged involvement, namely, Albert and Bernard simply stood by and watched, not participating in the emergency, simply observing, tutting, uttering a constant stream of unhelpful remarks. Worryingly, I noticed the patient's breathing appeared to be almost non-existent. Additionally, he appeared to have virtually no pulse. I informed everyone around, adding.

'I think we may have to perform CPR.'

'Right,' agreed Julie, the only other person getting actively involved with the situation. Suddenly, Albert started sucking profusely on his relatively new, dentures. I could not believe what happened next. Herbert had instilled into all of us during the First Aid course how we must reassure the patient, help them remain calm. For despite being unconscious, allegedly, the patient is still able to hear sounds around them. So, what does Albert say while sucking on his artificial, plastic teeth like a lozenge?

'That's it, he's a goner, he's had it. Sorry mate, but you're a goner, you're dead!' he adds, looking at the comatose resident lying on the reception floor hotel. This was Albert's parting shot to the poor sod lying prostrate and unconscious on the floor. Albert, while on his way back to dining table, obligingly informing and updating the other residents of the hotel, who had congregated and now peered inquisitively into the foyer, to where all the commotion was taking place. 'He's had it, he's snuffed it, he's a goner, he's dead.' After repeating his observations a few times, Albert immediately sat at the dining table and started tucking into his meal, which had just been put onto the table and carried on as if nothing had happened. I looked through the glass partition across at him, in complete and utter disbelief.

'Gee thanks Albert!!' I muttered somewhat contemptuously. What was even more unbelievable, at almost the same instance Albert departed the scene, Bernard began shouting after him.

'Albert, where are you going?' He immediately followed his colleague, attempting, unconvincingly, I might add, to exude the appearance of trying to catch his colleague, and return him back to the scene of the emergency.

Bernard then went straight into the restaurant behind Albert. I looked at Julie and shook my head in complete amazement and disbelief and tutted, while, at the same time rolling my eyes, in my head, displaying my disgust at the two of them. They had both dragged me into this situation and now left me *holding the baby.*

The main thing now, we had to try and revive the poor resident lying on the floor.

'Right!' I said to Julie, 'We'll have to get him over onto his back and perform CPR. Do you agree?'

Julie nodded her head in agreement. We rolled the dead, extremely heavy resident onto his back and pulled his airway open. I remained uncertain as whether Julie or myself would perform the operation, when thankfully, much to my relief, and no doubt Julie's, the Paramedics suddenly appeared on the scene. Just in the nick of time, like the seventh cavalry in a John Wayne movie. Julie explained what had happened to one of the paramedics while his colleague quickly provided an oxygen bottle and mask. Within a short time, the resident slowly revived. However, as a precaution, the paramedics took him to the nearest hospital. Julie thanked me for my help and I returned to my meal already on the table and now quite cold. In all fairness, the staff in the restaurant immediately took my cold meal away providing me with a hot replacement.

Meanwhile, Albert and Bernard had both finished their meals and carried on as if nothing had happened. My main thought being, I hope I didn't have a heart attack when only Albert or Bernard happened to be around. The other side of the coin being Ioan. He kept saying about the patient.

'He's drunk, that's all. He's been on the sauce all day!' Although, how Ioan knew that fact, is beyond me as we had all been at the pharmaceutical company all day, and nowhere near the hotel.

The next morning, while we were all having breakfast, the resident returned from the local hospital and went to pay his bill. It transpired he had completely underestimated the hotel bill that previous evening, not realizing how expensive the numerous bottles of Châteaux Neuf du Pape he had consuming were after putting a tab for the bottles of vino on his room bill. Ioan's cynical assumption proved to be correct in that respect, the resident had become intoxicated after drinking all day at a wedding, where he had been a guest, and began hyperventilating on being presented with his immense hotel bill.

That next morning, upon his return from the hospital, the hotel staff in revenge for the previous evening's events, took every form of currency he had on his person to pay for the outstanding bill, before finally sending him on his way back to Germany, leaving him without any money on his person whatsoever. It was yet another event involving Albert during his time at Chemkiln. From a personal aspect, I felt quite good, having kept reasonably calm during the incident. I felt I acquitted myself quite well. From then onwards, all this helped whenever I stayed at the Marina hotel, I became guaranteed prompt service at the bar or restaurant due to my limited First Aid intervention.

Since that incident, I have never really had cause to

use my first aid training to any extent, apart from minor cuts to fingers and such like, that is.

I just hoped in the future, I never suffered a heart attack with only Albert around to save me, or not save me, which would probably be the most likely outcome. Thankfully, it never happened, and I never experienced the opportunity to find out.

CHAPTER 7

The first significant project for which I became the designated Project Manager, occurred in the early spring of 1997. It would be the first of many such large projects in which I participated during my tenure with Chemkiln. This particular assignment involved working in Isleworth Hospital, a large NHS facility, located a short distance down the road from the newly established Sky television studios. The complexity of the project necessitated being away from home during the working week, with the team, including myself, staying in one of the cheap, local hotels in the vicinity. We did no work during the weekend apart from extreme emergencies or at the behest of the customer. Hence, for the duration of most of the project, we travelled home every Friday evening, returning early the following Monday morning, resuming where we left off.

A major pharmaceutical company had been carrying out research in the ancient hospital laboratories located in some of the hospital buildings. By mutual agreement, between the large medical conglomerate and the hospital authorities, the two agreed the drug company should relocate to a state of the art, purpose-built research facility located in Heston near the M4 corridor, situated a few miles west of their current location. The large pharmaceutical company desperately needed modern, spacious, well-equipped, state of the art research facilities, while the hospital urgently required more ward space to accommodate the substantially increasing patient numbers. The move thus had a symbiotic relevance with mutually beneficial consequences for both organisations.

A substantial portion of the project involved the movement of some hazardous, toxic chemicals. Both parties considered it prudent to contract the work out to a company dealing with toxic materials on a regular basis. Both the drug company and the hospital authority approached Chemkiln, suggesting the company submit a tender for the work. After putting forward their proposals and costs, much to their surprise, Chemkiln won the contract, putting me in charge of the project at the coal face, metaphorically speaking. My remit, to ensure the work was carried out safely, legally and, more importantly, as far as my employer was concerned, on schedule and well within budget, helping augment the company coffers.

During the initial stages, the project progressed rapidly. Unfortunately, after a couple of weeks; the work began slowly grinding to a halt. The manager in charge of the undertaking for the pharmaceutical company turned out to be a great vacillator, unable to make any on the spot decisions, forever conferring and seeking advice from his superiors, before finally committing himself to a ruling. He displayed an inability to *think on his feet.* As the project became more complex, the delays subsequently increased. Quite often, the team and I would pass many hours comfortably ensconced in the hospital canteen, consuming copious amounts of tea or coffee, awaiting some pronouncement or other by the Project Manager. An undertaking which should have lasted a few weeks, inescapably and relentlessly extended itself into months.

I previously worked on a project with my former employer Hyperwaste, with much the same scenario. That case involved a land reclamation project, which also materialized into a protracted, long drawn out affair. Chemkiln upper management did not mind the lack of progress, after all, they were still being paid an exorbitant daily rate by the customer, so did not find themselves out of pocket. The only problem which the General Services department experienced tended to be the continual pencilling in and amending people for future work or projects. It appeared to me, as an outsider to the pharmaceutical business, the companies encapsulated within that business seemed to possess a bottomless pit, whenever it concerned their finances.

One day, while sitting in the canteen awaiting some verdict or other from the Project Manager, a surgeon based at the hospital, asked if we would oblige and do him a favour by transferring some hazardous material from one of the laboratories, now designated to become a ward. He wanted the materials moved to an alternative location. We reported primarily to the pharmaceutical company, but had instructions, circumstances permitting, to carry out work for the hospital authorities. That day, time passed by slowly, it would be nice to have something to do, simply to help hasten the day along and occupy our time. Because of the delays and inactivity, the days became interminably lengthy, seeming to last forever. The other members of the team were busy bagging up old, confidential paperwork for incineration. I informed the surgeon that Bernard and I would perform the task. Besides, it would be good PR on behalf of our company.

We both followed the surgeon into a small room, substantially garlanded with numerous spiders' webs. The surgeon pointed to the several layers of shelves which snaked and looped around the off-white, dirty, walls. Secreted amongst the spiders' netting, the shelves appeared to be cluttered with dirty, opaque plastic bags of varying sizes. As far as Bernard and I could ascertain, the plastic bags appeared to contain lumps of dark brown, almost black bits, and what appeared to be lumps of rubber, immersed in a clear liquid. In most of the plastic bags, the once clear liquid, now had small bits of the encased material floating inside, souring the once transparent liquid, turning it the dirty, brown colour, we now saw.

After first sweeping aside the cobwebs from the plastic bags, barely legible labels, precariously secured onto the plastic containers, began slowly revealing themselves. Upon closer scrutiny, it became evident, from the faded descriptions on the labels, those bits of rubber, were actually an assortment of human organs; kidney, heart, liver, pancreas and such like. All used in the research to help discover a cure against the relentless, never-ending scourge of cancer.

I looked at Bernard who appeared to be growing paler and extremely pasty faced before my very eyes. His previously ruddy complexion, now ashen, as the blood, evidently drained rapidly from his rugged, extensively lined, well lived-in features.

'I know what we'll do Bernard,' I said, taking pity in his obvious discomfort.

I'll pass the bags down to you. You can pack them in the chem-boxes.'

He approved of my suggestion, which did seem the far safer option. Rather than the alternative of Bernard, while perilously perched on the stepladder, suddenly keeling over, falling the few feet, either onto the floor, or worse still, from my point of view at any rate, propelling his substantial frame on top of me.

We recommenced the tricky operation; Bernard, his feet firmly planted on the ground, while I gradually worked my way along, and up the shelves, passing the delicate bags containing their precious contents to my colleague. Meanwhile, he desperately averted his eyes, attempting to place the plastic bags blindly into the chemical boxes, and specially designed for the holding and transporting of toxic materials. The work progressed reasonably well, and my colleague's face appeared to be slowly regaining its normal, former ruddy complexion. I kept talking, trying to divert Bernard's mind, and stop him dwelling on the numerous body parts continually being passed to him.

'You can't tell these are parts of human organs, they almost resemble bits of rubber.' I continued, imparting my observations, whilst at the same time, exuding a nonchalant, devil-may care, sort of attitude and levity towards the whole undertaking. Bernard did not reply, and, as far as I could establish, nodded mildly in agreement.

By this juncture, I had progressed to the top shelf and grabbed another bag. This bag lay in waiting above my head, and I could not observe it properly. All I knew, it appeared to be much more substantial and heavier than the ones I previously handled. Cautiously, I pulled the plastic bag down from the shelf, its weight felt considerable. When in front of me, it became obvious, the large bag actually contained a human brain, minus the outer bony skull, which once encased and protected it. It surprised me how heavy the human brain is.

Unthinkingly, and without considering the ultimate consequences, I showed the gory contents of the bag to my colleague.

'You can tell what this is though.' I said with more than a hint of victorious perspicacity.

My colleague gazed up leisurely, only to observe the plastic bag together with its gory contents, the innards of a human head. Unfortunately, and almost immediately upon perceiving the grisly contents, the grey, ashen appearance once again returned, to Bernard's craggy features. Straightaway, Bernard made a speedy exit from the small room, hurtling his immense frame towards the nearest toilet, where he promptly regurgitated the immense breakfast he had consumed earlier that morning. The contents of his stomach, by this time, thoroughly mixed with the numerous cups of tea he had ingested earlier that day, making the contents of his stomach less viscous and more fluid and easier to project.

It all goes to show the idiosyncrasies, foibles and absurdity of life. Here was Bernard, a six-foot two inches tall, seventeen stone, former rugby player. By all accounts, an absolute vicious, competitive animal on the rugby pitch, now violently throwing-up at the sight of a human body part. For an instant, it was even touch and go as to whether he would throw-up or pass out.

I followed him to the toilets, which, fortunately for Bernard, were situated nearby. After firstly ensuring my subordinate was in no danger, I returned to the small room, and continued alone with the job in hand.

Bernard then spent a substantial amount of time on his knees, his immense arms entwined around the toilet bowl, with all the appearance of making wild passionate love to it. By the time he believed the contents of his stomach had been suitably evacuated, Bernard somewhat reluctantly, returned to the small room. By this time, I had completed the desired task. As he approached me, Bernard looked extremely embarrassed at what had transpired. A sort of defence mechanism cut in, he complained, quite bitterly, that some time prior to his rapid exit, I had slightly damaged one of the plastic bags. Subsequently, the formaldehyde fumes had affected him, causing him to evacuate his stomach in such a dramatic way. Desperately, trying to camouflage and mask the irrefutable fact, he was squeamish at the sight of human body parts, and which severely dented the macho image he liked to portray. He ultimately tried to salvage some of his damaged ego, by generating this story.

Choosing to ignore his protestations and complaints, I instructed him to go and get some fresh air, and then have a highly sugared cup of tea, while I transported the chem-boxes, which I had loaded onto a trolley. Then I'd transfer them to the room which the surgeon had asked for them to be relocated.

I came to the designated room and knocked on the

entrance. A young female assistant opened the swing doors; her face covered by a surgical mask, covering her head, and a surgical hat which completely shrouded her hair. She also sported a lab coat which shielded her working clothes. On her hands, she wore a pair of thin latex gloves.

'I have some samples I was told to transfer here.' I explained to the beautiful, piercing blue eyes which gazed back at me above the surgical mask.

'Where shall I put them?' I enquired.

'Just put them over there,' she instructed, while at the same time blithely pointing to a vacant corner of the room.

I pulled the trolley loaded with the chem-boxes to the assigned location. The room contained four other people, three of whom were also attired in the same garb as the young female commissionaire who had let me in. However, one of them stood by a table. It was the surgeon who had requested the transfer of the body parts, with his back to me, dressed, as he had been earlier; white shirt, blue tie, and black trousers. He wore no lab coat, or any head gear, and as far as I could ascertain from my viewpoint, no surgical mask either. The only concession he made towards his protection, a pair of thin latex gloves. Suddenly, he turned around to look at me, while at the same time stepping aside, allowing me an unrestricted view of what he had been so engrossed with.

'Do ya fancy opening a butcher's shop? I have some meat for you here?' he jested, in his barely comprehensible, broad Glaswegian accent.

I looked to where his index finger lamely pointed. There on a wooden board, neatly pegged out, profuse lengths of human intestine, which the surgeon, seconds earlier, had been meticulously securing, with the aid of small pins onto the large wooden board. I also perceived the board to be liberally smattered with blood from the same intestines.

To my astonishment, and my own personal gratification, the site of all the blood and gore completely unfazed me. I had overcome my aversion to blood. I had come a long way, since my embarrassing episode of passing out in front of Herbert during the First Aid Course.

'Nah,' I retorted, 'No profit to be made in it. The major supermarkets have cornered all the business.'

I then asked. 'Shouldn't you be wearing a lab coat, or some sort of protective clothing?'

Before he could begin to answer, the young female assistant, who had earlier let me into the room, interrupted, answering on his behalf.

'Other surgeons do,' she answered, with more than a hint of vexation and exasperation in her tone, then appending to the statement.

'He should do the same!!'

The surgeon smiled, totally unconcerned and undaunted by his assistant's frustration and annoyance.

'Och, the others are not as careful or as good as me. Look, not one drop of blood on my clothes.'

With that, he did a twirl, for the benefit of all and sundry in the room, allowing everyone a complete view of his proudly uncontaminated clothing. He then carried on with his work, while the young assistant still glared angrily at him, with her piercing, disapproving, blue eyes.

The thought went through my mind at the time; it was just as well Bernard was not with me at that instant. I considered the possible consequences and outcome; my colleague, by this juncture, probably lying prostrate and comatose on the floor, having caught sight of all the human, blood-soaked intestines so neatly and precisely pinned to the wooden board.

The project lasted a few months, during that time, we all stayed in a hotel located in Hounslow, just a few miles away from Isleworth. For some inexplicable reason, the department secretary was unable to find accommodation nearer to the hospital. The Shalimar Hotel where we all lodged, possessed a certain charm and eccentricity, but unfortunately, was located just a few hundred feet beneath one of the aircraft flight paths to Heathrow Airport.

There was absolutely no need of an alarm clock or an early morning call. From dawn, or certainly some ridiculous hour in the morning, the huge intercontinental *red-eyes* began arriving from across the pond, landing at the international airport to discharge their passengers. Their arrival would be evident to all the residents in the hotel, briefly indicated by the violent rattling of the windows, and the reverberation of the building, as the jets screamed just a few hundred feet above the roof of the building. Often, the room started to shudder, and ornaments or any loose items in the room slowly began walking across the surface on which, seconds earlier they had been resting quietly. The worst, tended to be the arrival of the very first aircraft of the day, as it screamed overhead, rudely awakening even the heaviest of sleepers.

Every morning, I awoke this way, immediately jumping up, startled, confused and bemused. My gentle slumber, violently disturbed by jet engines screaming, as if in pain, hurtling along, just above the apex of the large building. Sleep was impossible after the first aircraft of the day made its approach. For, approximately every minute and twenty seconds, another bus of the airways made its way towards Heathrow Airport. I will never understand or comprehend anyone wanting to live in, or near Hounslow, with all those commercial jet aircraft screaming overhead. I was more used to a semi-rural existence. All this noise was, and still is, completely alien to me. Unfortunately, the worst culprit for emitting high decibels, and it pains me to admit this, appeared to be Concord.

Since the late sixties, I had been enthralled and captivated by the development of this feat of engineering, with its futuristic, sleek design. It was an aircraft I admired greatly. Regrettably, it did make a thunderous, ear shattering noise whenever it landed or took off from any airport. Nonetheless, this drawback did not diminish my admiration for the brilliant, magnificent, supersonic aircraft, apart from being directly under the flight path to Heathrow Airport, the Shalimar Hotel.

The manager of the hotel was in his early thirties and quite '*Jack the Lad.*' Always on the make, and someone Arthur Daley could most certainly have picked up a few tips from. He was always having arguments and disagreements with one of the female waitresses. Unfortunately for the waitress, she was in her mid-fifties. More than likely, at one time she may have been quite attractive, and a lure for any normal heterosexual male, but those days had long since gone, and life had worn her down, draining her of her vivacity, figure, beauty and, ultimately, her charm. If she possessed an hour glass figure and been in her twenties, then the Manager of the hotel may well have acted differently towards her. Unfortunately, she possessed none of those attributes, hence, the manager treated her with disdain, considering it not worth his while or effort in flirting or being agreeable towards her.

The upshot being, during our time there, he sacked her, probably using some trumped up reason as the basis for her discharge. There were ramifications following her dismissal. Despite being directly under the flight path to Heathrow, the hotel was quite a popular hotel and place to stay. Quite a few residents choosing to stay there, more than likely because of the low cost. Consequentially, breakfast in the restaurant tended to be a busy time, generating a fair amount of hustle and bustle. It was not a self-service buffet, requiring two waitresses. Unfortunately, the morning following the dismissal of the waitress, breakfast time was particularly busy, and the lone waitress found herself unable to cope with the volume of residents.

With everyone vying for her attention, often venting their feelings concerning the lack of service, the waitress finally lost her composure and temper. Standing on one of the few empty tables, she began making a speech concerning the dismissal of her colleague, and how the manager had not found a replacement, or arranged cover, leaving her to cope on her own. She then went into a tirade about the manager, complaining about how useless he was. She also, paradoxically, requested we all put in a complaint about the inadequacy of the service to the owners of the hotel, allowing her to then put the case forward on behalf of her colleague, as well as herself, where she could make the shortcomings and behaviour of her boss evident to the owners of the hotel. Because the manager was such a sleazebag, of such an unlikeable disposition, I wrote a letter of complaint that same evening. Besides I felt some sympathy for the hard-pressed staff. Other residents must have done the same, because the following week, the sacked waitress was reinstated

The project lasted far longer than planned or anticipated, extending into May. I recall, because of another memorable event at that time, the General Election of May 1st, 1997, in which the Labour Party under the leadership of Tony Blair gained political power by a resounding landslide majority.

Albert assisted with the project at the time. All that week, the two of us became involved in heated political arguments. With Albert being a staunch, unflinching, unquestioning Thatcherite, while I, on the other hand, completely disagreed with her and her successor's (John Major) political doctrine and dogma. During our time working on the project, we became embroiled in some heated discussions on various points of political policy.

The evening of the election, I watched events as they unfolded on the television screen in the privacy of my hotel room. I had only intended watching the initial results as they came in. But the results proved to be so unbelievable; I remained fixed to the BBC, watching the results for the constituencies, as, one by one, they came in. It appeared from the initial constituency results, the Labour Party were heading towards an historic landslide victory over the incumbent Tory government. I continued watching the television until the early hours of the morning, when there could be no doubt as to the eventual outcome, finally succumbing to a blissful sleep.

On entering the breakfast room that morning, Albert wore a despondent expression. After a prolonged silence, he eventually spoke. 'Well it looks like the sodding Labour Party has won... Heaven help us!!'

'Yeah bloody fantastic, isn't it?' I retorted while at the same time, exuding a broad smile, just adding to and exacerbating his annoyance.

All that day the television screens were filled with images of Tony and Cherie Blair smiling for the cameras, grinning and shaking hands with all and sundry. The music 'Things, Can Only Get Better,' the Howard Jones song, made popular by D-Ream, blasting out and adopted as the New Labour anthem, discharging out from the television every time the new Prime Minister appeared. Who could have foretold just ten years later, he would depart the office he so coveted, vilified by the media and public opinion. Such is often the fate and destiny of those who seek and attain public office.

A few weeks later the project at Isleworth hospital eventually ended, but it is yet another project with lasting memories, preserved in the old memory banks.

CHAPTER 8

Although, I suspect the reader would most probably not concur, I consider myself to be a bit of a romantic, and occasionally enjoy playing cupid, assisting in guiding true love along what I deem to be its correct and rightful path. An improbable and incongruous place for discovering true love in this instance proved to be the old fishing port of Grimsby, along with the seaside town of Cleethorpes. The two small towns located right next to each other on the coast of Lincolnshire.

Why Grimsby? You may well ask. Well, primarily because one of Chemkiln's major customers, a large pharmaceutical company called Novartis, had a giant research and manufacturing facility based there. This particular romantic episode revolves around Novartis. Chemkiln frequently sent a team of two people to the site, generally about once a month, to pack and remove the laboratory test samples and chemicals for incineration in the South Wales furnace.

The first time I turned up at the facility with Albert, resulted in a major cock up. Not of my making, I must emphasise. Being the first time for both Albert and I to go there, meant we both had to undergo an induction and safety course, before being allowed onto the huge site. The induction course included pointing out the site rules such as speed limits, as well as highlighting the major health and safety risks present on the facility. The course also pointed out the sounds of the alarms and sirens, what they indicated and which safety assembly points we should go to in case of emergency.

Chemkiln had only won the contract six months before I joined the company, hence the reason Albert had not been there before. Usually one of the other supervisors and another operator performed the necessary task. On this occasion, a few of the staff in the department had decided to take their vacations, leaving the section short staffed. So, it fell to Albert and me to perform the task. I must point out, Albert had no chemical qualifications, he did however have project management skills, that's where his forte lay, hence his reason for being a supervisor in the department. On this occasion, he acted as my assistant while I performed the chemical inspection and segregation. Usually, Rupert or another supervisor accompanied by one of the operators, Bernard Evans or Steve Williams performed the task.

The cock-up came about because Albert and I had not been booked into the induction course, which had to be booked at least twenty-four hours in advance. The two of us turned up at eleven thirty in the morning, fully expecting to attend the course scheduled for twelve noon, as we had been informed. Unfortunately, Jamie, Chemkiln's sales representative, had not booked us in. This meant we could not attend the session until the following morning.

Despite all our protestations, cajoling and pleading, the supervisor in charge of the course would not let us attend the course planned for that day. Rules were rules and not intended to be broken or superseded at any cost. He proved to be a real '*jobsworth,*' doing everything by the book, not willing to compromise. Subsequently, Albert and I found ourselves at a bit of a loose end, unable either to go onto the site or book into our hotel, the King's Royal, and our accommodation arranged for the next two nights of the project, located in Cleethorpes, a few miles along the coast from Grimsby. With it being noon, far too early to check into our rooms.

So, considering the situation and circumstances and unable to sign in to the hotel, we decided to pay a visit to the local fishing museum in Grimsby. The excursion proved to be quite edifying and interesting, giving an insight into the history of the fishing industry, not only in Grimsby, but throughout the UK, learning about Captain Birdseye. My irascible colleague appeared not to be as enamoured with the museum as myself, constantly criticizing and decrying the quaint, old, evocative establishment. After spending a couple of hours or so in the museum, we decided to have a coffee, before finally making our way to the hotel. Not bad; being paid by our employer to spend an afternoon strolling around an old fishing museum.

The next day we finally managed to attend the induction course, obtaining the necessary contractor's passports to enter the huge facility.

For the next few monthly visits to the Novartis site at Grimsby, I usually ended up with the job of being the supervisor. Instead of being accompanied by Albert, I generally ended up being assisted by Steve Williams. From the outset, it became evident Steve had a crush, or to be more precise, a sexual desire pertaining to one of the females, and our main point of contact. Her name, Cheryl Foot. Cheryl worked in the Environment Department of Novartis. Steve obviously wished to get to know her better, and I do mean in the biblical sense. Steve had been to the site on quite a few occasions before I eventually appeared on the scene. By this stage, he had built up quite a friendly rapport with Cheryl. Only a complete idiot could fail to recognize the sexual chemistry and tension which existed between the two of them.

At the time, Cheryl was in the process of getting a divorce from her husband. She was to all intents and purpose, virtually single and available. In addition, at this period in his life, Steve did not have a girlfriend or any romantic acquaintances. With Steve being about the same age as the attractive Cheryl, both in their early thirties, it could become a union pre-ordained by heaven, along with a little help from myself, of course. Unfortunately, with both parties being shy and reserved, neither one of them appeared to possess the courage, or impetus to make the first move.

So, on my third visit to Novartis with Steve, I had decided enough was enough, and these two would-be lovers must be somehow brought together for their own good and well-being. From the side lines, the whole situation manifested itself as an exasperating, painful, frustrating experience to observe. Hence, the reason I allocated myself to be the person to expedite the romantic process and speed this mating game along. For our final evening of this visit to Novartis, I invited Cheryl to have a meal with us in one of the restaurants nestling along the bracing Cleethorpes seafront, near to where she lived. Mercifully, she agreed, and a pleasant evening transpired in one of the local steak houses. Although, I did feel a bit like a spare prick at a wedding, while both cautious, possible lovers flirted, all be it outrageously, but still maintaining a certain modicum of caution with each other.

During the evening, I suggested, without any pretence of finesse, subtlety or tact, Cheryl should invite us back to her place for a coffee. She readily agreed, without exhibiting any obvious hesitation or reluctance. I must confess, her quick acceptance did take me by surprise. After first consuming a few drinks in a couple of the nearby pubs, the three of us finally headed back to her abode, and the promised free cup of coffee. My plan appeared to be coming together. I anticipated, after the two bashful, would-be, possible lovers had consumed some alcoholic drinks, their inhibitions would be suitably reduced and, wham, bam, thank you ma'am. I hope the reader will excuse the rather vulgar, course expression? After quickly drinking or rather gulping down my mug of coffee at Cheryl's, and almost burning my mouth in the process, I told the two of them I felt tired, as well as having a bit of a queasy stomach. Informing them I would make my own way back to the King's Royal Hotel for a relatively early night. Of course, a complete fabrication, and excuse on my part. Steve felt obligated to accompany me back to the hotel. I quickly cut him short. Stopping him, before instructing him to stay at the house, and have a chat with Cheryl, along with another coffee. Reassuring him I would be fine, and able to make my own way back to the hotel without his assistance. I told him I would see him at breakfast at about eight the following morning. So, I headed down the quiet street, leaving the night's events to unfold as they may.

As agreed, the next morning, at around eight o'clock, I made my way to the breakfast room of the King's Royal Hotel, expecting either to find Steve already sitting there, or for him to appear shortly after me. I ordered a full English breakfast, along with all the trimmings, then waited. Eight fifteen, no Steve, eight thirty, no Steve. Eight forty-five, still no sign of Steve. The hall clock chimed nine o'clock. As if on cue, at that precise instance, Steve shuffled through the main lobby of the Hotel. He looked dreadful, with dark bags under his eyes; supported by extremely bloodshot eyeballs which looked like an AA road map of Europe, with all the associated miscellaneous road colours. Not only that, his normally well-groomed hair appeared in complete disarray, distinctly unkempt, protruding in all sorts of directions, exuding an unrestrained, revolutionary life of its own.

Yet, despite his distinctly exhausted appearance, Steve exhibited a broad smile, just like the cat that got the cream. He conveyed the unmistakable air of a man who had been subjected to an exhaustive, but highly memorable night, of substantial, libidinous, energetic, sexual activity. I, on the other hand felt pleased for him and, like a man who had fulfilled his objective, experienced pride in myself. A self-congratulatory thought passed through my brain, 'My work is done.' The laconic Steve proved to be not very forthcoming about the previous night's events, portraying his true gentlemanly style and philosophy. Eventually, all I managed to extract from him, the brief underrated statement, 'Well, that turned out to be a very good night.' When I questioned Cheryl about that night, on my next visit to Novartis, a different scenario prevailed on her part. For, she proved far more forthcoming and distinctly less reticent, or inhibited about the night's events.

'Yes, that was the first time we slept together and had sex; although, neither of us got much sleep that night.' While imparting this information, the attractive Cheryl gave me a sly smile, together with a cheeky wink. She displayed absolutely no embarrassment concerning that night's sexual events whatsoever. From then on, due to my obvious helping hand in changing her life for the better, we became good friends.

From that eventful night onwards, the romance between Steve and Cheryl blossomed unremittingly. Most free weekends, either Steve travelled to Cleethorpes or vice versa, Cheryl travelled to Caerleon, where Steve lived on his own.

Unfortunately, things don't happen without some sort of consequence. Newton's Third Law of Motion kicks in; 'to every action there is an equivalent and opposite reaction.' In the case of this happy, heart-warming story, I received the backlash, and an opposite reaction from Damien Phipps.

Damien and Steve had been friends since their schooldays, hence the main reason the two of them now worked in the same department. When he became elevated to the head of General Services, Damien asked Steve to come and work for him. Not only that, Steve appeared to be the only person Damien went drinking with. Nobody liked going for a drink with Damien, particularly whenever there happened to be a football or rugby match being shown on the large television screen in a pub. Damien could be quite loud, raucous and vociferous, particularly whenever some incident occurred during a sporting fixture. Shouting and cursing, loudly and profusely at the screen. His barbed, profane comments being mainly aimed at the referee or other match officials.

By no stretch of the imagination, could Damien ever be considered a quiet or reserved sort of person. Being with him in a pub usually ended being a 'cringeworthy, I wish I were dead', thoroughly embarrassing experience. We have all been there, watching a rugby or football match in a pub with a loud, boisterous friend, who appears to be the only voice to be heard above the general cacophony of sound within a crammed room. All you want to do is crawl under the nearest table, and pretend that person is not actually with you, but belonging to someone else in the pub. Damien generally tended to be that individual you wished to completely disown.

Unfortunately for Steve, he lived quite close to Damien. Steve usually ended up accompanying him to the pub for any large sporting event. He felt both obligated and threatened whenever Damien approached him to go to the pub, because of all the reasons previously mentioned. So, I appear on the scene in Chemkiln. Within a short space of time, Steve is no longer available to go drinking on the weekends because of his new, highly physical relationship with Cheryl. Often Damien approached me in the office during the early days of their relationship, only to enquire if there was indeed a relationship between Steve and Cheryl, or, to be more precise, extolling his particular form of the Queen's English and political correctness, 'Is he shagging her?' Obviously, not happy with the state of affairs

It soon became evident and common knowledge as the relationship continued to develop exponentially between the two of them. Within a year, Cheryl relocated to Caerleon, and Damien lost his only true drinking buddy. I received the fault for all this upheaval; but felt unashamedly guilty as charged, but extremely proud of my part in the burgeoning relationship. Although, I like to think, I simply became the catalyst for the blossoming relationship. For, in all probability, it would have eventually come about anyway. I just gave it an ever so gentle prod in the right direction. Damien always resented the part I played in the events concerning Steve and Cheryl. But, I must confess, this fact did not worry me one iota. Steve and Cheryl's love life and the part which I played in it, was none of his business, despite him being head of department.

I ended my time working for Chemkiln some time ago, but now, years on, the last I heard, Cheryl and Steve are still living happily together. My initial appraisal of the situation between the two of them proving to be spot on. Pity I could not have been so prescient concerning my own love life in the past.

CHAPTER 9

Each project performed during my time working for Chemkiln, had its own peculiarities. Characteristics with certain memories and charms, prevalent only to them. It was precisely those individual, distinctive recollections which added to the allure and excitement of the job, which is why I enjoyed working for Chemkiln so much. Every one of us working for the department, had no idea what each task would bring up, or the difficulties and adventures we may possibly encounter during the tenure of the various individual assignments.

Surreal incidents often occurred, such as the time I had to inspect some chemicals earmarked for disposal at an electronics company located in Galashiels Scotland. My main point of contact at the company was an engineer called Steve, a Londoner who had relocated to the region. As I was on my own for a few days, he offered to accompany me to the town, and introduce me to some of his friends. It was March 17th, Saint Patrick's Day. That evening, we toddled off into Galashiels, to frequent the pubs. One particular pub had strong Irish connections and celebrated the Patron Saint of Ireland in an exuberant and flamboyant style, completely decking the pub out in green. I had a fantastic evening, me, a Welshman, Steve's friends were Scottish, he was English, and we were all drinking in a Scottish pub, surrounded by loads of Irishmen, most of whom appeared to be dressed as Leprechauns celebrating their Patron Saint. It was a totally surreal experience, but it did feel like we were all part of a truly United Kingdom.

Going to BAT (British American Tobacco), based in Southampton, also proved to be an edifying experience. Whereas, most large employers in the nineties, voluntarily began imposing a no-smoking policy on their sites, BAT bucked the trend, actively encouraging smoking on their premises. In fact, I think it was virtually compulsory for their employees to smoke. Visiting the canteen proved to be an interesting experience, with numerous cigarette machines in the canteen and ashtrays on all of the tables. Along the corridors there were also cigarette machines providing Benson and Hedges, one of the main brands manufactured by BAT, which adorned the passageways like paintings. Of course, legislation eventually put an end to this situation, with a no-smoking policy implemented in all workplaces, rightly so.

Another memorable project which I recall fondly, involved *'exploding eyeballs.'* The project was identified as such in all the correspondence and files relating to this particular venture. Throughout the many years of its existence, the famous Moorefield Eye Hospital in London, in its pursuit and advancement of medical science, had unsurprisingly, managed to accumulate a considerable number of eyes and miscellaneous parts of those organs, as a consequence of its research and development, in the pursuit and advancement of medical science. The ancient and by now, surplus organs were amassed together in glass jars and immersed in water. Nitrocellulose was then added to the contents as the preservative.

During an inventory and inspection, it was noted some of the jars and their contents had dried out as a result of insufficient sealing on some of the lids screwed on top of the jars. This inadequate sealing allowed the water to slowly evaporate, leaving a residue of nitrocellulose on the eyeballs, as well as the tops of the jars and inside the glass containers.

Nitrocellulose had once been used extensively in the manufacture of the old film reels of the twenties and thirties. During storage, the reels frequently spontaneously combusted, causing the warehouse facilities to burn down, resulting in the loss of priceless, irreplaceable films, made during those infant years of the motion picture industry. The managers at the eye hospital knew the dangers of dealing with nitrocellulose, but they also realised if the chemical remained dissolved in copious amounts of water, the risk of the eyeballs spontaneously combusting was minimal. However, with the water having evaporated from a considerable number of the glass vessels, they became concerned those eyeballs could literally explode, possibly resulting in a fire at the ancient hospital. Hence, the reason for our company's involvement and the bizarre name assigned to the project. The remit of the company; to inspect all the jars and put water back into any desiccated containers. The jars with the low water content had to be first immersed in a container of water, the jars then gingerly unscrewed, allowing the water to percolate into the glass container and re-dissolve the chemical. Following this operation, all of the old jars and eyeballs were to be later transported to our incinerator and destroyed.

No-one knew if the unscrewing of the dry jars would actually cause the chemical to explode or possibly spontaneously burst into flame. This being the reason for the immersion in water during the delicate operation. Once completely re-filled with water, all the glass jars were placed into fifty litre plastic Mausers, then completely packed in vermiculite. The vermiculite, an inert material, prevented the glass jars rubbing together, and possibly breaking. Once filled, the Mausers were sealed with a plastic lid, securely located in place using a metal clamp. The drums then finally labelled with the correct UN identification for transportation.

It had been decided Bernard Evans would accompany me to perform the task. The others in the department were unaware and ignorant of Bernard's attitude towards blood, body parts, organs and such like. I was not about to 'dob him in,' following my experience with him at Isleworth Hospital, for it could possibly result in him losing his job within the department.

The first day of the project involved travelling from South Wales. Then, upon arrival, setting up and sorting all the equipment. Following an overnight stay in the centre of London, the second day of the project began in earnest. Because of the incident concerning Bernard at Isleworth Hospital, I considered it prudent, that perhaps I should be the person performing the delicate operation with the glass jars, while Bernard packed them away in the plastic Mausers, then completely fill the plastic drums with vermiculite, ensuring the brittle glass containers did not rub up against each other and possibly shatter during the journey by van.

The first few sample jars we came across, appeared to be completely full of water, requiring no action to be performed on them. Each jar full of water was gingerly placed in the first plastic Mauser. However, after a while, I eventually came across some glass jars which obviously contained no water at all, with a discernible white residue of the potentially dangerous nitrocellulose preservative evident within the confines of the containers. So, suitably safely attired and with much trepidation, I commenced the exercise of immersing the first jar into the large container of water, gingerly unscrewing the lid. Initially, it was difficult to ascertain when the thread of the lid was at the end, and I could pull the top away easily from the glass jar. The next few jars, the lid suddenly parted company with the thread of the container, allowing it to fill with water.

Regrettably, because of this sudden parting of company between the lid and the glass jar, some of the eyeballs managed to escape. The released eyeballs began bobbing about in the water contained in the large plastic vessel. I tried keeping Bernard ignorant of the fact, but lamentably, not well enough, because, he managed inadvertently to catch sight of some of the bobbing eyeballs staring vacantly back at him from the surface of the water. They frequently, teasingly disappeared, before resurfacing from beneath the liquid. It was déjà vu and a repeat performance of the occurrence at Isleworth hospital involving the exposed human brain, some months previous. Yet again, Bernard's face became ashen and pale, just as it had during the episode at the Hospital. He made a rapid dash for the nearby toilets, where he once again indulged in his seemingly, newly-acquired pastime of projectile vomiting into the nearest ceramic basin. From then on, I made certain Bernard no longer caught a glimpse of any eyeballs which accidentally escaped from the glass jar, then started bobbing about in the immersion water. I am additionally rather inclined to believe, Bernard also made certain he did not look in my direction, while I performed the delicate operation, for fear of a repeat performance with the contents of his stomach.

We eventually managed to complete the project in the stipulated two days, without any further exhibitions by Bernard on how to violently regurgitate the contents of one's stomach in the fastest time possible. But unfortunately, before completion of the project, Bernard did have another incident; this time completely unrelated to the actual project but involved the upper set of his false teeth.

At the time of the *exploding eyeballs* project, Bernard was in his fifties. During his earlier and much younger years, Bernard had been an enthusiastic rugby player; his usual position involved participating in the engine house of the team, with the pack of forwards. Because of his enthusiasm and fire, he had managed to become separated from most of his teeth. In those early, amateur days of rugby, there were no gum shields. Health and Safety most definitely did not apply, with only wimps using such protective equipment as gum shields. The adage held sway, 'men were real men in those days.' The general accumulated loss of his teeth was either because of over exuberance on Bernard's part in his playing on the field, usually due to over-zealous tackling. Although, generally, the higher percentage of his teeth loss resulted because of some numerous violent altercations with miscellaneous players from the opposing teams.

As previously stated, Bernard could be described as an absolute animal on the rugby field, with his competitive spirit completely enveloping and dominating his psyche. Yet, off the rugby pitch, one could not wish for a more temperate, placid, amiable fellow. Now in the autumn of his life, the competitive spirit, he possessed had unfortunately presented him with a complete lack of natural choppers, forcing him to chew his food with the aid of synthetic false teeth.

On the first night of the *exploding eyeballs* project, Bernard had set his mind upon devouring a chocolate Yorkie bar, which he had left in the freezer to stop it melting during the warm, sultry day. Bernard had no way of knowing, but the setting for the fridge compartment had been adjusted by the previous occupant of the room, to an extremely low setting. By the end of the day, the Yorkie bar was so cold and hard, it could have been employed as an attachment for a diamond drill bit and utilised for oil exploration in the outer regions of the North Sea. All afternoon, Bernard had been eagerly anticipating chomping through his chocolate bar. He unravelled the wrapper from the chocolate bar with delight and relished the sight before him before ultimately biting ravenously into it.

Unfortunately, with the Yorkie bar being so solid and sturdy, it immediately broke the top set of his dentures. The false teeth were unable to compete with the tensile strength of the frozen chocolate bar. It was most decidedly, a totally uneven contest, terminating with the dentures surrendering in utter despair, splitting right down the middle.

Bernard came to my hotel room explaining the turn of events and explaining the predicament. He now found himself in. Before this juncture in time, I had never seen my colleague without his top set of gnashers in situ and discovered myself unable to prevent myself from laughing. He looked so risible, endeavouring to talk through his upper gums, spraying saliva in all directions. I felt no guilt in laughing at his predicament, for I had not one iota of doubt, he too would have done the same had the tables been turned. Following a rather soggy discussion, it was decided to try and reconstitute the top denture with the aid of non-hazardous adhesive. I went to the nearest shop and managed to purchase some innocuous adhesive material. After taking it back to Bernard's room, we began the delicate operation of reassembling the useless dentures. Bernard performed the delicate operation and took it upon himself to use one of the luxurious fluffy towels supplied by the hotel to hold the broken dentures in place while he applied the adhesive. I must admit, he performed the operation with a fair amount of dexterity and manipulation considering the enormous size of his hands which appeared to be the size of large dinner plates.

At last, Bernard completed the task of sticking and reconciling the two separated sections of his top set of dentures together. He kept the dentures on the luxuriant towel allowing the adhesive to do its work. After what Bernard considered to be a reasonable period, he decided to restore the dentures to their rightful location, on the roof of his mouth. Unfortunately, much to Bernard's chagrin, a large amount of the material from the towel, refused to dislodge itself from the adhesive, and in the process remaining firmly stuck to his dentures.

Despite Bernard's frantic attempts at completely removing the material, a significant amount remained attached to the false teeth. Eventually, he admitted defeat, deciding to put the repaired dentures plus the material from the towel into his mouth. His contorted face was picture to behold after he had carefully located the dentures in his mouth, he could barely talk. Through his muffled voice I was able to discern after ignoring the words of profanity, he likened the whole experience was like having an expensive shag pile carpet shoved in his mouth. Throughout the next day, all he did was complain vociferously, albeit slightly muffled due to his dentures being coated with the woollen material from the hotel towels. The dentures and wool firmly entwined.

Another memorable project concerned Sussex University located in Brighton. Unfortunately, it was discovered one of the large electrical transformers on the campus contained PCB, the banned, toxic transformer oil. Chemkiln had won the contract to firstly remove the oil by transferring it into barrels, and then completely remove the large electrical item for shipment and disposal at our incineration facility in South Wales. The barrels of PCB and the carcass of the transformer to be transported on the same articulated flatbed vehicle. After the University maintenance department and electricity board completely electrically isolated the transformers, in the first two days of the project Steve Williams, Ioan Winston and I pumped the PCB from the transformers into two hundred litre drums, and then flushed the transformer with base oil to remove the last traces of the toxic material. Once the base oil was completely pumped out and the carcasses of the transformer, empty, we began undoing any bolts securing the electrical equipment to the ground, allowing it to be lifted freely and loaded onto the flatbed of an articulated vehicle.

The last day of the project would simply involve loading the drums containing the PCB and the transformer onto the articulated vehicle. Dead simple. At most, we estimated the whole exercise would take no more than an hour, to an hour and a half to perform. Unfortunately, the best laid plans of mice and men etc. etc.

I had organised everything for that last morning, the vehicle to transport the material to the incinerator, a substantial crane to lift the transformer, onto the flatbed. The maintenance department of the university had agreed to let us use their site fork for lifting the pallets loaded with the ten drums onto the flatbed. Everything had been arranged for nine o'clock that morning, although the team and I arrived early at eight fifteen, to prepare things, being proactive as we thought, anticipating an early departure. At eight forty-five, the driver from our regular haulage company appeared at the location on foot and minus his vehicle. The driver's nick name was 'Bunny.' One can only hazard a guess the origin of this name; my hypothesis invariably involved it being associated with sex in some way.

I asked him where he had parked his vehicle. He casually informed me that he had left it in the main University car park, having no wish to drive a forty foot around a large site he had no knowledge of, and suddenly discovering he had driven his vehicle down a cul de sac with no easy means of extricating it. I had no problem with that, considering it a very sensible decision. After discovering our location, he assured me he would return with his vehicle and be back at our location within ten to fifteen minutes.

The crane had arrived earlier and was in position at five minutes to nine, ready to lift the transformer onto the flatbed. By nine fifteen we were all awaiting Bunny and his vehicle. The transformer had to be loaded onto the front of the flatbed, its weight being absorbed by the rear axle of the front motor unit. The drums containing the PCB from inside the transformer were on pallets, ready to be loaded onto the flatbed, behind the transformer.

Nine thirty arrived, still no sign of Bunny and his vital articulated vehicle. Another thirty minutes passed. By this time, I decided to contact his main depot and discover Bunny's whereabouts. Bunny had not provided me with his mobile number, I had thought there was no need; after all, he would not be too far away. His manager was very apologetic. It seemed a rather irate and apoplectic Bunny had been in contact with him. Apparently, during the short interval of time Bunny had taken to find our location, and return to his vehicle in the main car park, he discovered the students arriving for their morning lectures had selfishly and inconsiderately parked their old bangers around his articulated vehicle, utterly enveloping and surrounding it, completely blocking the vehicle in. Bunny was in the process of trying to contact the numerous miscreants and free his vehicle from its unwanted and unexpected entrapment, hence the reason for the delay in his return.

Also, I was not amused, the team and I had anticipated an early departure, now we had to wait patiently because of some inconsiderate, selfish students.

Bunny finally appeared with his vehicle at quarter past eleven, two and half hours from when we last saw him. And when he emerged from the cab of his unit, he was literally not a happy Bunny. (Sorry, I must apologise for that, but I found it impossible to resist.).

The air was blue, with loud profanities from Bunny interspersed with quiet mutterings from the irate driver.

'Stupid bastards! They are supposed to be the cream

and intelligentsia of society, I wouldn't trust them to make me a cup of tea. The stupid twats!!'

Bunny carried on with his necessary tasks, preparing his vehicle for loading, while at the same time, continually berating the alleged intelligence of the student community, which he considered to lay be at the lower end of the IQ measurement scale. Not only castigating Sussex University but the whole of the student population in general. No person in higher education was spared from Bunny's vitriolic effusions. After all, he too had been anticipating an early departure, now completely decimated by a few inconsiderate, selfish students.

As estimated, the actual task of loading the drums together with the carcass of the transformer took less than an hour to complete plus another half an hour by the time I had the necessary legal hazardous waste paperwork signed off by the customer. So, within two hours of Bunny finally turning up with his vehicle, we were ready to leave and departed in convoy at about one o'clock.

Being a Friday afternoon, the journey from Brighton to Newport took the team about five hours including a lunch stop; eventually I arrived home at about seven that evening. God only knows what time Bunny finally arrived home after delivering the contents of his vehicle to the incineration facility.

We just never knew what could transpire with each project and the unforeseen circumstances which could arise.

CHAPTER 10

Milton Keynes, the new, purpose-built town, is famous, or some would maintain infamous, for its plethora of roundabouts together with its neat, geometrically arranged streets and highways. It is also renowned for possessing concrete cows, being home to the Open University and Bletchley Park. The latter being the location for the World War II code breakers, often referred to as 'Ultra' by Winston Churchill, because he considered it to be his ultra-secret weapon in his fight against Hitler and the Nazis. However, Milton Keynes is not the sort of place one normally considers to be twinned with Sodom and Gomorrah; conveying an exhilarating, exotic, debauched life, but instead, quite the opposite, that of being a town famed for its distinct privation of a colourful, wanton social life. A place considered to be instilled with a certain serene tranquillity. However, I can assure the reader, as with most large conurbations, there are always places of disrepute to frequent, usually surreptitiously hidden away. The secret is to discover where such places are ultimately located. Seek and ye shall find. Milton Keynes proving to be no exception to this premise.

Hoechst Rousell, another pharmaceutical client which used Chemkiln, hired the General Services department to completely decontaminate its Research facility at Milton Keynes. The work included the usual tasks; removal of all hazardous chemicals for incineration at our facility. Also, decontamination and disposal of laboratory equipment and furniture, likewise, condemned for destruction in our state-of-the-art incinerator.

I had been allocated the position of Project Manager for this assignment, usually accompanied by one or two of the regular team members; usually Bernard Evans, Ioan Winston or Steve Williams. Chemkiln sometimes hired two or three men from another contracting company to help the project run smoothly. The project lasted for many weeks, with the numbers in the team varying throughout, dependent upon how well the project appeared to be progressing, utilising the additional contractors as a form of labour safety valve, using their services as and when required.

The team often stayed at a place called the Pear Tree Lodge, situated near the customer's facility. The hotel possessed a large, well provisioned bar, a nice décor, offering nice cuisine, with an ambience which projected an agreeable, friendly atmosphere. One evening, the hotel management organised a karaoke night. Now Stella, my partner, had introduced me to this particular form of cheap, social entertainment while vacationing one summer in Vilamoura, a plush Portuguese resort situated on the Algarve coastline. It turned out to be an introduction she later came to regret as I became a karaoke freak, primarily because of this initial acquaintance to the phenomena.

Now isn't it a sort of truism, the people who are the most tone deaf and unable to sing, invariably, are the very ones who are more than likely to indulge in this cheap form of entertainment? Unfortunately, I fall into this category, particularly after having consumed copious amounts of any alcoholic beverage. I must confess to being unable to sing, even if my life depended upon it. What transpired this particular evening proving the point beyond any reasonable doubt. There were five of us still involved with the project, Ioan, three subcontractors and me.

We had all devoured a wonderful meal in the hotel restaurant, washed down by a fair amount of alcohol, mostly beer. After consuming the meal, we all retired to the bar for the evening, where we discovered the aforementioned karaoke evening in progress. While carefully perusing the karaoke song book, listing all the tracks in the repertoire on offer, I observed my favourite song; Roy Orbison's 'Pretty Woman.' The alcohol coursing through my veins conferring me the necessary courage to complete the form, then hand it in. But first, I needed to return to my hotel room and retrieve a pair of sunglasses. The sunglasses, I felt, were crucial to enhance my forthcoming performance, being an essential, intrinsic part of the act. Ioan looked at me bemused while I quickly departed the bar in order to obtain the indispensable accessories necessary for my forthcoming performance. I quickly returned with my dark spectacles, then handed the already completed song form to the Disc Jockey.

'Where did you disappear to?' enquired Ioan, inquisitively. I showed him the sunglasses, then smiling, informed him.

'My props, I need them to perform the song!'

He looked back at me, still exuding a thoroughly bemused, perplexed expression.

As if by magic, the bar suddenly began filling up with people. Within a brief period of time, I guessed there must have been close to sixty people packed into the small room. At last, the DJ called out my name.

'Vinson Chard will now sing the Roy Orbison classic, Pretty Woman!'

Nervously, I approached the stage. The alcohol pumping around my system helping slightly negate any nerves which I experienced at the time. After first going through a routine of theatrically donning my sunglasses, I started singing the classic, timeless song, as the words slowly appeared highlighted on the television screen. Unfortunately, through the passage of time and much use, my cheap dark lenses had become badly scratched, causing me great difficulty in reading the words on the screen. Periodically, I kept lifting my sunglasses in order to read the words, trying to ensure my singing synchronised with the music, unfortunately, failing miserably in my endeavours. My singing ended up being completely out of tune, and totally out of sync, sounding entirely flat. Overall, the performance ended up being a complete debacle and fiasco. From my point of view on the stage, the audience appeared cocooned in complete darkness, the effect of the darkness being exacerbated by the tint in my badly damaged sunglasses, making it almost impossible to perceive the audience. Finally, my purgatory ended, experiencing immense relief at being able to vacate the stage, after the abysmal, embarrassing exhibition of how not to sing, but how to make a complete and utter prat of oneself.

I walked off the stage, upon entering the darkened section, then removing my defective sunglasses, I realized the bar had almost completely emptied, apart, that is, from my colleagues together with a few of the stalwart regulars still firmly ensconced at the bar. Everyone else had departed the room, whether as a result of my diabolical performance or because the people had en masse decided to go elsewhere, I have no idea. All I remember is the landlord asking, or to be more precise, pleading with me not to sing again. A request I acquiesced to. That was the beginning and, as it transpired, the end of my brief karaoke career in Milton Keynes.

A few weeks later, with the end of the project in sight, only my colleagues from Chemkiln and myself remained, with no additional contractors. The team decided to go downtown Milton Keynes into the main town and sample the night life on offer. Altogether, there was Ioan, Steve, Bernard, Paul Morgan, one of the other supervisors who decided to stop by, and myself.

We all piled into a cab, having decided to partake of a meal in the main town instead of the hotel. This particular night, we chose Mexican, having seen this specific restaurant during one of our previous infrequent visits into the centre. During the journey, we quizzed the taxi driver concerning the best places to go for an evening of decadence and utter debauchery. He told us of a pub/club called the George, and which every Thursday night had female strippers as the main entertainment. The driver also recommended a couple of other nightclubs worth visiting after the show. We all agreed; first on the itinerary, a spicy Mexican meal, then to the George to observe some naked female flesh. Finally, to complete the evening of profligacy, a visit to one of the night clubs recommended by our chatty, friendly taxi driver.

After consuming a fantastic meal in the restaurant, we made our way to the recommended den of iniquity. Paying our entrance fees, we all entered, only to discover the downstairs of the establishment to be completely packed with lecherous, lascivious men. We all made our way upstairs to the balcony bar, which overlooked the stage, on which we assumed the nubile female strippers would shortly perform. Finally, the master of ceremonies jumped sprightly onto the small platform, informing his salivating audience, for that evening, the George had booked two female strippers. He immediately introduced the first of the duo, a blonde with a magnificent figure. She began dancing, or to be more precise, gyrating provocatively around, putting on quite a performance. Dancing around the stage and amongst the downstairs audience. We all began regretting having made the decision to view the show from the first-floor balcony.

Finally, the blonde girl completed her act, then went backstage. The next performer to appear turned out to be a brunette. The first girl had been attractive with a nice figure. But this second artiste was Aphrodite, a goddess; truly beautiful with dark, almost black, curly hair. She slowly began divesting her attire, exhibiting a magnificent pair of pert, large, well-shaped breasts.

I became transfixed upon this gorgeous vision and perfect example of the naked female form. She was absolutely gorgeous, with an amazing figure. My eyes must have been on stalks as they surveyed this wonderful, naked female apparition gyrating in a sexually enticing manner around the rostrum. The young woman danced erotically around for some time, before eventually grabbing one of the young men from the audience, then pulling him onto the small platform, which passed for a stage. The young man gave a lascivious smile to his male friends, who looked on enviously as this vision of feminine beauty danced seductively around their compatriot. Slowly, the young woman began undoing the buttons on the willing participant's shirt, while at the same time exuding immense sexual provocation, rubbing thick cream onto his naked chest.

She continued provocatively dancing around the willing participant. Slowly bending him forward over the table, strategically located on the platform. The young man kept looking back and smiling at his friends, obviously anticipating something wonderful and erotic was about to happen. While he bent over the table, the young, nubile exotic dancer slowly pulled down the willing participant's trousers, together with his underpants, exhibiting his naked posterior to all and sundry. She continued seductively dancing around before picking up a large aerosol can, full of cream. Slowly, deliberately, and with exotic skill, she began spraying copious amounts of the cream onto her young, alacritous volunteer's naked back. The next moment, she danced enticingly off the stage where she immediately picked up a large whip, which had been surreptitiously propped up in the corner, after which she returned to the stage holding on tightly to the lethal implement, in her right hand. In an instant, she pulled the whip back as far as she possibly could. Suddenly, this Aphrodite, this vision of loveliness and female perfection, began exuding the extreme maniacal look of a sadist. It seemed with obvious pleasure, she brought the whip down with as much venomous force as she could muster. As if making some sort of demonstrative statement, articulating her hatred for all men, the poor, unfortunate individual on the stage being the recipient for all this seemingly pent-up aggression. Suddenly, everything I observed seemed to manifest itself in slow motion. The tip of the Marquis de Sade's implement of torture, came down upon the large mound of viscous cream, which projected upwards on the young man's back. Almost the instant the whip connected with the white mass of cream, it sprayed everywhere. Making the onomatopoeic 'splat,' 'splat,' 'splat,' sound as globules of the thick, viscid cream hit the bemused audience on various parts of their clothing and bodies.

Even the bouncers on the door, suddenly discovered their faces covered in the viscous cream, after the material had projected itself across the room at high velocity; much to their extreme annoyance and obvious irritation. The front members of the audience also experiencing the same ignominy. Just like a giant white fan unfolding, the viscous material radiated outwards and horizontally. Its epicentre being the young man's back. Instantaneously, the young, luckless man shot upright, while at the same time, emitting an involuntary, blood-curdling scream. An indication of the excruciating pain he suddenly underwent. As he stood upright, the remnants of the cream on his back, slowly draining into his open, exposed trousers, which lay around his ankles. Meanwhile, my colleagues and I remained safely shielded on the upstairs balcony and not subjected to the indignity of the whipped cream covering our faces and clothes. Only the audience on the lower floor suffered this humiliation.

Thankfully, the cream only appeared to spray horizontally and not vertically. Ioan laughed at loudly at all this chaos, bedlam and mayhem taking place beneath us. I just looked on in disbelief, at this vision of feminine pulchritude and beauty, who had maliciously and with obvious intent caused all the pandemonium, confusion and disruption, directly beneath us. Before making her rapid exit, the exotic dancer picked up her discarded clothing, and quickly vacated the scene, but only after cheekily curtsying to the dishevelled, confused audience in front of her.

The show over, everyone gradually vacated the premises. The bouncers still had smatterings of cream over their faces and clothing, which helped mitigate somewhat their aggressive, pugilistic appearance, not making them seem so terrifying or frightening. My colleagues and I headed for one of the nightclubs recommended to us by our knowledgeable, informative taxi driver. Later that evening, through the flashing strobe lights, I caught sight of the young man who had been the unfortunate victim of the striptease artiste's attack. I observed him rubbing the base of his spine where the whip had connected so violently and unexpectedly with his body. Evidently, he still felt some pain. I wondered what imaginative explanation he would later relate to his wife or girlfriend concerning the prominent red mark, so evident on the back of his pale torso. I felt so grateful I had not been the one to have been subjected to the torture inflicted by that beautiful, yet blatantly, malicious young woman.

I have an interest in knowing where sayings originate; such as freeze the balls off a brass monkey, the whole nine yards etc. The first, incidentally, relates to piles of frozen cannonballs, not testicles. The second refers to belts of machine gun bullets on the Spitfires during the second World and twenty-seven feet long (nine yards) when the pilots had used up all their bullets, they had gone the whole nine yards. Now there is the famous saying, '*Cock and Bull Story.*' I often wondered the origins of this wonderful saying.

During one of the weeks working at Hoechst in Milton Keynes, we were unable to get rooms in our normal hotel, The Pear tree, which meant we had to go further afield to obtain accommodation. Eventually ending up in one of the old villages, situated on the outskirts of Milton Keynes, called Stoney Stratford.

There are two Hotels right next to each other one called 'The Cock, 'the other called 'The Bull.' with a narrow alleyway running between them. Apparently, so the story goes, centuries ago, there were two old washerwomen. One worked in 'The Cock,' the other, in 'The Bull.' The two old women regularly chatted and gossiped to each other from the upstairs rooms across the narrow alleyway. The women became well known locally, and gradually the term, '*Cock and Bull Story*' came into being. To my immense joy, my colleagues and I stayed in the Bull for a couple of evenings.

So, summing up, despite Milton Keynes exuding a rather boring reputation for its lack of atmosphere and excitement to the outside world, I thoroughly enjoyed my time spent working there. Speaking from personal experience, I can dispute the erroneous personality which Milton Keynes portrays to the outside world.

Perhaps it is twinned with Sodom and Gomorrah after all?

CHAPTER 11

During the latter part of the nineties, because of my deteriorating relationship with Stella, the physical side of our partnership diminished rapidly, due to reasons I will shortly come to. However, ironically, my love life would increase exponentially with other willing female partners, thanks mainly to a number of factors such as; the personal home computer, the fledgling internet, and the internet provider America On-line (AOL). The substantial amount of time I spent away from home working for Chemkiln, adding to the mix of events and opportunities. With all the above conditions somehow managing to align themselves at the same time. This alignment, some may even define it as kismet or serendipity, would ultimately alter my path through life, greatly amending the road ahead of me.

Having acquired a personal computer and modem, my access to the internet became complete, opening up a whole new world, which blissfully exposed itself to me. The youngsters of the new millennium with their high-speed broadband, 4G, WI-FI, I-pads and instant access to the net, cannot possibly comprehend how long it took to access the internet during its infancy and initial development during the nineties. Millennials would have been utterly appalled the time it took in those early days to actually access the internet. The 'dialling up' process unfortunately meant the noisy accompaniment of annoying rasping, screeching noises which emanated from the modem, as it desperately attempted to enter the vast world of the internet, via the antiquated, totally inefficient telephone lines available at the time. I will refrain from using the term electronic super highway, for the reasons just mentioned. In the nineties, the internet highway could in no way possibly be ascribed with the adjective, 'super.'

After eventually becoming a member of the internet

fraternity, it had not been my specific intention to embark upon sexual liaisons, proposing only to use the internet for broadening my intellect and its entertainment value. That is, until one night, after consuming a vast amount of alcohol; Stella confessed, with vitriolic candour, how she hated my company, finding me boring, bland and totally uninteresting. She went into detail explaining about the buzz and excitement which she initially experienced during the early days of our relationship and our endeavours attempting to keep our relationship a secret from everyone, primarily Cindy, my partner, and Stella's fiancé Neil. Stella stated those days of thrills and excitement had long since evaporated. I also recalled those early days of the affair, our secret meetings, and the ever-present danger of being discovered.

Our sexual liaison remained undiscovered by all our work colleagues at Repeat Controls. Stella drunkenly affirmed, the fact our liaison remained a secret, enhanced and magnified the whole situation, adding to the buzz, thrill and excitement of the affair. Now, a few years down the line, our formerly clandestine, furtive relationship was no longer a secret, known to all. We had become an accepted couple, settling into a normal, dull, mundane domesticity, with its regular routine and existence. Stella continued explaining drunkenly, how she no longer experienced a high, buzz or thrill concerning our partnership. Throughout the conversation, Stella made it abundantly clear, through continual rote and repetition, how she considered our life, or more specifically, her life, to be monotonous, tedious and dreary, taking on an evening job as a barmaid, explaining the job meant she could spend as little time as possible with me. As she put it, '*Meet far more interesting and exciting people.*'

However, the language she used, distinctly more profane and abusive. The following day, I sat an extremely hung-over Stella down on the sofa, and referred to the previous evening's discussion. Now sober, Stella claimed no knowledge of the previous evening's events, maintaining the onset of alcoholic amnesia, due to the vast amount of Vodka she had ingested. She argued, almost pleaded, she did not really feel that way about our relationship. However, as I pointed out to her, the ancient Romans often recited the dictum, '*Vino Veritas,*' (in wine [there is] truth). What she had said the previous evening had hurt me immensely, cutting me to the quick. I also believed it to be how she truly felt about us, and our relationship. Vodka, not wine, had simply loosened her tongue, as the ancient Romans pontificated two thousand years earlier. Perceptive people those ancient Romans.

From that night onwards, as far as the opposite sex was concerned, I considered myself to be a '*free agent.*' The AOL chat rooms providing ample opportunity to become acquainted with willing, sexually aware and experienced females on-line. And boy, did I meet quite a few. Working for Chemkiln, made it far too easy in adding a day here, and a day there, onto my time working away, utilising the black hole of time to indulge in illicit assignations.

Often, either the woman or I would not feel the desired magical, biological, or chemical attraction towards each other. On frequent occasions, very little transpired, apart from some minor petting. Sometimes we had sex, but often, it usually ended up being a one-off encounter, making no further arrangements to see each other again. This suggestion between us being implicit.

Then, one memorable evening, I accidentally began chatting to a woman named Rebecca, in one of the many AOL chat rooms available at that time. Rebecca dramatically affected my life for the next eighteen months. She was a forty-something divorcee, living with her nineteen-year-old daughter in Bognor Regis. Rebecca had a well-paid job as a legal advisor, employed by a large company, and she was very intelligent. She possessed a degree in law. Her particular expertise, dealing specifically with consumer law. She sometimes appeared on terrestrial television, giving legal advice and opinions, acting as a pundit for a few consumer programmes, especially on the BBC, mostly on daytime television.

Our initial meetings on the internet, during my infrequent time spent at home, began innocuously enough with the usual chit-chat. Gradually, the messaging between us became more risqué. We sent photos of each other. It was Rebecca who first introduced me to the concept and wonders of cybersex. The first time we purposely indulged in this experience took place on a Saturday night while Stella was at work and a few weeks after we first met on-line. It was quite an experience. To such an extent, immediately afterwards, she gave me her telephone number and we decided to communicate the old-fashioned way, verbally via the telephone.

Her voice sounded so husky, sexy and alluring. Once again, I became aroused despite our previous indulgence in cyber-sex. We began talking as if we had known each other for years, not weeks. Rebecca complimented me on my Welsh accent. I complimented her on her sexy, ever so refined, accent. I made Rebecca aware of my circumstances with Stella. Whether she completely believed the situation or not, I can only guess at. The photographs indicated her to be an attractive brunette, with a voluptuous figure, not like one of the extremely thin, super models, often erroneously referred to as the *ideal woman*. I much prefer women to have some flesh on their bones, and not able to scrutinize the rib cage poking through their outer skin, resembling some emaciated prisoner of war.

Our telephone conversation went on for a while; we obviously had a lot in common. It was then we both decided we must meet, literally in the flesh. The following week, I went into work and perused through my allocated weekly work schedule for the next month or so. From the schedule, I had to somehow manufacture a meeting with this sexy female lawyer from the south coast of England. For a couple of weeks, a meeting was impossible. My schedule involved working in the far North of England, making the prospect of meeting someone from Bognor Regis, impracticable. Then a couple of weeks later, the ideal scenario came up. I had to work away in the Midlands until Thursday, but then spend the next day, a Friday, in the office.

I put it to her, if Rebecca was agreeable, perhaps we could possibly meet in Reading or Swindon on the Thursday night. On the Friday morning, I would be able to drive straight from the hotel to the office in Wales, and Rebecca drive directly to her place of work on the south coast. I could lie to Stella, telling her I would be away working on a project until the Friday evening. Not that she even cared all that much.

At first, Rebecca appeared to be agreeable to the idea, and seemed quite enthusiastic at the prospect of us spending the night together. However, when I contacted her nearer the date, after making all the final arrangements, she appeared to be experiencing cold feet.

'Vinson?' she enquired, nervously down the phone.

'I know we said we would book a double room in the hotel. Would you mind if we did not go straight to bed, allowing us first to become better acquainted?'

She then added, 'Also, promise me if I don't feel attracted to you, is it okay if we just sleep in the same bed?'

It sounded fine to me. After all, as I had previously

discovered, these sorts of things can cut both ways. She might not find me attractive, but then again, I might not find her sexually appealing either.

We arranged a secret tryst for the Thursday night, during a February, for seven thirty pm in the car park of the IBIS Hotel Swindon. I would travel from the Midlands. Rebecca would travel straight from her place of work in Portsmouth.

The day of the rendezvous arrived, and all I could think about was the forthcoming meeting, and the evening ahead with Rebecca, which held a lot of promise. We completed our work mid-morning, then journeyed back to South Wales. First, I headed to the incinerator plant, to drop off all equipment. I then drove straight to Swindon, arriving at the car park of the IBIS hotel at seven, where I waited nervously for Rebecca to appear. We both knew each other's car, the model, colour and registration which we had divulged during our discussions on the internet.

For the first half hour, I walked around the car park, checking the make, model and registration of the cars, ensuring Rebecca had not arrived before me. I must have appeared very suspicious on the CCTV cameras, strategically positioned around the car park.

The allocated time of seven thirty came and went by, then seven forty-five. I began to envision Rebecca not appearing, or not coming, in whatever sense of the word. Suddenly, a dark blue Ford Fiesta, exhibiting the car registration number I had been given, entered the vast car park, then pulled into a vacant parking bay. After a brief interlude, the female driver emerged. I looked from the confines of my new Rover 75, and appraised the woman alighting from the Ford Motor Vehicle. My God, she looked gorgeous, far more attractive than I had hoped. With her collar length, neatly maintained, brunette hair, attired in an expensive long, dark woollen coat, she looked very classy, exuding a certain demure *je ne sais quoi*. This elegant, stunning apparition stared into the darkness; obviously desperately seeking my car, which she believed to be amongst the throng of parked vehicles, resting serenely in their individual space on the tarmac. So, this was Rebecca. I emerged from my car, but only after first flashing my headlights, just like one of the characters in a John le Carré spy novel. When I thought she was looking directly at me, I waved. She acknowledged me, collected her travel bag from the boot of the car, then walked seductively in my direction. While she was doing this, I collected my travelling bag from the back seat of my vehicle, then slowly walked towards her.

As Rebecca approached, she appeared to become even more alluring and attractive. There was no way this woman was going to make love to me tonight. She was way, way out of my league, and would indubitably back out at the last minute. I was punching well above my weight. But, to my astonishment, upon coming into closer proximity, she began exhibiting a broad smile, showing her perfect white teeth, indicating no obvious signs of disappointment with her date. From my initial appraisal of the situation, things looked distinctly promising. Peering directly into my eyes, she also probably discerned no dissatisfaction on my part.

'Vinson, I hope?' she enquired rhetorically.

'Rebecca?' I replied inanely.

She pecked me gently on my left cheek, in an almost formal, sisterly fashion.

'Shall we go to our room first and drop off our bags?' she enquired rhetorically.

We both went to the reception desk, and then asked for the room, which I had booked under the name of Chard. There was no problem with Stella checking and finding out about the evening, even in the unlikely event she ever decided to scrutinize my credit card statements. During the previous year, I had signed into so many hotels with my work, there was absolutely no way she could possibly keep tabs on my movements by examining my extensive credit card statements.

The reserved hotel room was on the third floor. We entered the lift, slowly ascended then walked along the corridor. I unlocked the door allowing Rebecca to enter first. We both surveyed, the void, your typical hotel room, with a double bed. After putting my bag down, I approached Rebecca, looking down at her, before finally speaking. She was five feet five inches to my five feet ten inches.

'At last, here we are alone,' I said.

'Oh yes, 'she replied sexily, then, putting her arms around my neck, pulled my head towards hers before kissing me passionately. It was a far, far different kiss from the first one she gave me in the car park moments earlier. This was most definitely not a sisterly kiss.

At the same time, she pressed her body tightly against mine. She most certainly, must have perceived my immense pleasure and delight at meeting her.

I needed this to stop immediately, or I would not be responsible for my actions. Reluctantly, I pulled away. I had promised her we would have a meal so that we could become better acquainted. I was not about to renege on my sincerely made pledge. I thought I detected a slight hint of disappointment in her face, but, at that instant, remained uncertain about that belief.

'I promised you a meal, and a meal you shall have. It will also give us an opportunity to become better acquainted, as we agreed.'

Rebecca nodded in agreement.

We both went down to the hotel lobby, perusing the menu on offer in the hotel restaurant. The meals provided by the hotel, appeared pretentious, overpriced and, as is generally the case with those types of meals, miniscule: your stereotypical, ostentatious, nouvelle-cuisine. Immediately afterwards, you require a kebab or some sort of fast food, to fill the immense void still left in your stomach.

'This sophisticated woman will probably want to eat here' I thought to myself.

Imagine my pleasant surprise when, after first carefully scrutinising the menu, Rebecca looked at me and spoke.

'Well it all looks rather showy, pretentious and overpriced to me. All I fancy is good honest pub fare. Steak, salad and chips would be my ideal meal, and would satisfy me.'

More and more, I was warming to this woman. She was just too good to be true, attractive, sexy, classy, and now, I discovered, down to earth, and apparently entirely devoid of any pretensions.

I kept wishing in my mind, *'God I want you and most definitely want to make love later.'*

I went to the reception, asking the girl at the desk to book a taxi. Acting uncharacteristically like an experienced man of the world, which Rebecca seemed to find appealing and pleasing.

Within five minutes the taxi arrived. I showed off by opening the door for my date. Getting into the cab, I asked the driver to take us to the sort of pub which provided good, honest steak meals.

The driver, in all fairness, took us to a nice pub, not too far away, near the centre of Swindon.

As neither of us was driving we both had wine and later spirits. The evening was superb; the pub had a wonderful ambience and oldie, world character. The waitresses who served us were exceptionally friendly. The steak meals were cooked to perfection, with generous, reasonably priced portions. Everything neatly clicking together like a jigsaw, all enhancing the magic of the evening.

From the outset, Rebecca and I hit it off. We talked about our lives, our jobs, families, and our likes and our dislikes, films and music, covering an immensely eclectic range of subjects. Unsurprisingly, the time passed by as a blur, and seemed like a nanosecond. I had most certainly enjoyed the evening spending time in the company of this attractive, sexy, witty, charming lady, even If we did not later physically consummate our friendship later.

The waitress came around, enquiring if we required any coffee. Rebecca sneakily peeked at her expensive watch. The time was about ten fifteen. She immediately seemed to take on the mantle of Cinderella, wishing to vacate the premises as quickly as possible. Without any form of embarrassment Rebecca spoke to the waitress.

'Could we have the bill please, and would it be possible to book a taxi to the IBIS Hotel?'

'Certainly, I will arrange all that,' the waitress replied courteously. Recalling, at the same time, I observed her exuding the hint of a knowing smile.

After the waitress had gone, my date grabbed hold of my hand across the table, squeezing it tightly. She looked at me seductively and spoke.

'I've had a fantastic evening, but I think it is time we went back to the hotel to be on our own, don't you?'

She appeared to be indicating what I had hoped for all evening. If Rebecca was not serious, then she had to be the ultimate teaser of all time.

After paying the bill and leaving a generous tip, it seemed the arrival of the taxi appeared to take forever, although, it was, probably, only a few minutes. Additionally, the journey back to the hotel turned out to be sheer purgatory, I wanted to jump on the bones of this woman, who kept talking inane nonsense to the female cab driver. I could not wait to be alone with her in the hotel room. Rebecca, not the cab driver that is. Rebecca later told me, she kept jabbering to the driver to take her mind from contemplating pouncing on me. A couple of times I attempted to place her hand on my knee, but she resisted, fearing the consequences should she comply. She later confessed, at that instance, her thoughts conjured up the following headlines.

MIDDLE AGED COUPLE ARRESTED FOR HAVING SEX IN THE BACK OF A SWINDON CAB.

Or possibly, in the Sun.

COITUS IN A CAB!

After exiting the taxi, we both rushed up to the hotel room, quickly closing the door behind us. I teased her.

'I don't suppose you want to make love after that big meal then?' I jested.

'What do you think?' She replied, once again, exuding a provocative, seductive tone in her voice.

Before I had chance to utter another word, Rebecca pulled me to her and kissed me as if she wanted to eat me whole. Despite the meal consumed earlier. She must have been a Praying Mantis in a previous life. Before I knew it, we were grappling on the bed, impatiently ripping each other's clothes off. There is nothing which is quite as erotic as that first time of mutual physical surrender between lovers, saturated in all the physical tension, anticipation, atmosphere and sexual excitement being generated. Rebecca had magnificent breasts which appeared to almost flow surreptitiously and gently out of her bra as she unclipped it. She was obviously aroused as indicated by her erect nipples. I knew the foreplay Rebecca enjoyed from our cyber-sex discussions, hence I knew all the right moves to make and what she enjoyed having done to her. After kissing, caressing and exploring each other's bodies all over, we finally consummated our relationship, culminating in energetic, passionate lovemaking on top of the bed clothes.

As is generally the case with first time lovers, it was wonderful. We both lay on the bed naked, perspiration running down both our bodies. Rebecca looked gorgeous, lying there naked as the day she was born, broad, satiated grins exuding from both our faces.

'Wow!' I said, 'That was fantastic. However, I do feel extremely bad about taking your virginity.'

She laughed out loudly at my attempted joke.

'Sorry to disappoint you there lover, but that went a long, long time ago.'

And it was my turn to laugh out loudly.

Eventually, I suggested I needed a shower to cool down and wash the perspiration from my body.

'May I join you?' she enquired looking at me, fluttering her long eye lashes, deliberately, looking seductively directly at me through her sexy, invitingly warm, brown eyes.

We both went into the shower which was nice and warm, not too hot. We then lathered each other's body in more than one sense of the word. Almost immediately, we began French kissing and indulging in tongue gymnastics. Then I suddenly stopped and smiled at her.

'You want to make love here, now. You do realize we have a Health and Safety Issue. Statistics show the amount of slip related accidents which take place in the shower and bathroom each year are phenomenal?' I enquired.

Rebecca looked up at me and once again gave me her wonderful smile.

'Well, I'm willing to take the risk if you are?'

We resumed our passionate kissing and made love, this time standing up in the shower, while the warm water cascading over our aroused bodies.

Afterwards, returning to the bed we both lay under the sheets exhausted and satiated. It was then Rebecca confessed earlier she had not wanted to go for a meal and would have preferred a pack of sandwiches from the local garage, regretting making me promise to take her out first, desiring to spend more time alone in the room. Rebecca also declared she had wanted to make love from the first instant of our encounter in the hotel car park informing me she said she had experienced a sort of lust at first sight. Indeed, so had I. Thinking about it, had we spent more time in the hotel room, I probably would have succumbed to a heart attack from over exertion. The lady proved to be insatiable.

We both slept for a few hours before Rebecca woke me in the early hours, arousing me by gently kissing me down my naked body before finally indulging in her wonderfully adept foreplay techniques. Once again, we succumbed to the ultimate pleasure.

We both left the hotel around six thirty in the morning allowing us both to arrive at our respective work places at a reasonable time. I must have looked like death when I sat down at my desk in the General Services office. But it was the wonderful, exhilarating feeling of exhaustion, the sort of exhaustion one experiences after an amazing night of ardent, zealous lovemaking.

This relationship continued for the next eighteen months we met in numerous hotels and locations; Bristol. The Severn Bridge hotel, Cardiff, Swindon and I frequently went down to her home in Bognor Regis while her daughter stayed at her boyfriend's.

I had miraculously managed to keep this entire secret from my work colleagues and Stella, as if she even cared anyway. It was easy to keep it from my work colleagues as I always did it when we were not always tagging the stay with Rebecca onto my work calendar. However, one time we had a waste project in The States of Jersey, removing waste for incineration. The work meant staying on the island for a couple of weeks. By this time into our relationship, Rebecca and I talked frequently via the phone.

Rebecca knew I was staying in Jersey, then mid-week, completely out of the blue, she casually informed me she had been looking into flights for the following Friday afternoon from Eastleigh, near Southampton to St Helier. She had never been to Jersey despite living so near to it on the south coast. I thought she was joking about staying for the weekend. Especially when she informed me flights were about £120 return. She jokingly informed me, I don't believe you are worth £120'

A couple of hours later, I received another phone call. This time she informed me

'I have investigated another airline, part of KLM. They can do a return flight Friday returning early Monday morning, in time for me to go to work, all for the bargain price of £65. Now, I do believe you are most definitely worth £65, so I will be arriving in Jersey four thirty pm on Friday. Which Hotel are you staying in?'

I was ambivalent about her suggestion. Yeah, sure I wanted to spend the weekend with her, but no, I did not want my work colleagues to become aware of our affair. Not that they would say much, but they sure as hell would give me a lot ribbing about it, and certainly make my life hell in the months ahead.

That Friday, we changed hotels. The one we were currently staying in was fully booked for the weekend, so we had to relocate to a Hotel on the sea front of St Helier. I nervously informed Rebecca of the situation and the name of the hotel, also telling her to wait in the lounge after she reached the hotel; we should be there about five thirty.

A quick phone call was made to the Pilot hotel requesting if my single room could be amended to a double as my wife would be staying the weekend. The receptionist told me there was no problem.

Now came the difficult part, explaining to Bernard that I would have female company, and not Stella, all that coming weekend.

He was thrown off guard and glared at me.

'You dirty, randy sod, what am I supposed to do all weekend?'

'I'll give you some of the float. I am certain you will find some things to amuse yourself for a couple of days.' I replied.

Friday afternoon arrived, and Rebecca had told me everything had been booked and she would wait in the hotel lounge as agreed.'

Bernard and I finished work at five pm then headed for the hotel. Carrying our bags to reception we asked about our reservations. The receptionist smiled at me and then caught me off balance, totally embarrassing me in the process.

'Your wife.' The receptionist deliberately emphasised the word *wife*, indicating she knew full well that Rebecca was most definitely not my wife.

'Your *wife* has already gone up to your room and has signed in. It's room twenty-one on the second floor'

Later, I discovered Rebecca had signed in under her own name, and not Chard. A dead give-away from the outset.

Bernard just looked at me and tutting and shaking his head incredulously.

The two young female receptionists at the desk just smiled at me, my face crimson with embarrassment, I hastily made my way to the room. I wanted to get away from everyone at the earliest opportunity.

Quickly making my way to the room, I knocked on the door. Rebecca opened it. She was wearing a silk dressing gown. Meanwhile, in her right hand, she proffered me a full fluted shaped glass of sparkling champagne.

'Hello lover.' she said, smiling at me before kissing me passionately on the lips, then handing me the effervescent, alcoholic beverage.

'I've run a nice hot bath; I thought we might have one together. You should see the bath, there's plenty of room for two, it's immense.'

I smiled back at her.

'Well you certainly embarrassed me downstairs with Bernard and the two receptionists,' I replied.

'Don't worry, they are used to it. Just relax; a nice hot bath along with other things I have planned hopefully will ease all that pent-up tension away.' She retorted nonchalantly.

With that, she began undressing me slowly, until I stood in front of her completely naked, and to her satisfaction, evidently ready for action. Rebecca loosened the sash of her dressing gown and let it fall from her body; we were then both naked, and in an instant, kissing passionately.

The steaming, frothy, soapy, fragrant bath beckoned. I sat behind her and she sat between my open legs as I sponged her back, while at the same time gently kissing the nape of her neck. With our glasses of champagne on the side, we both periodically sipped the revitalising sparkling wine. There was no haste to make love, unlike our first encounter. We had all the weekend, with both of us intent on savouring every glorious moment. No rushing things this time. Besides, making love in a bath tends to be a tricky, messy business, generating a mini tsunami, the result, as well as the sexual climax, usually resulting in a completely saturated bathroom floor.

After removing the grime of the day from my body in the bath, we both transferred to the bed, firstly drying ourselves off, before eventually finishing what we had initiated in the immense bathtub.

That weekend was memorable. I took the company vehicle, showing her the sights of Jersey. Despite having lived on the south coast all her life and near the Channel Islands; before that weekend Rebecca had never visited the Islands. During the weekend, we went for meals and walked through the complex maize of streets which is St Helier. She wanted to meet Bernard, but I had no intention of that happening. I never knew what Bernard would say or do to embarrass me. He could be a bit of loose cannon in that respect. I therefore considered it a much wiser option to keep the two of them far apart.

During that enjoyable weekend, we even went to see the film, 'You Have Mail' about a couple; Meg Ryan and Tom Hanks who meet via the internet, which we considered exactly similar to our situation.

Unfortunately, that weekend passed by far too quickly, and Monday morning quickly arrived, when Rebecca had to catch her flight back to Eastleigh on the south coast.

After that weekend, we had a few more assignations, but it was those few days spent in Jersey which will always be indelibly etched into my memory. Unfortunately, within a year, Rebecca phoned to inform me she wanted much more from our relationship and, as far as she was concerned, extremely unhappy with our too infrequent meetings. I was devastated but knew in my head she was absolutely right. I lived in South Wales, with a well-paid job and career, travelling around the country. She lived on the south coast, and had a lucrative career. With that fateful telephone call, our lust driven relationship came to a sudden and abrupt end. *C'est la vie, c'est l'amour*. But what a sexual adventure, what memories, what exquisite fun!!

CHAPTER 12

Ayot St. Lawrence is a small, quaint, picturesque village, secluded and surreptitiously tucked away in the beautiful Hertfordshire countryside. It is also where George Bernard Shaw, the famous Irish playwright, wit and raconteur, finally ended his days. I must confess, until commencing work for my new employer, I had never heard of the small hamlet, but that was one of the many sundry benefits derived whilst working for a national hazardous waste company, with many of its major customers scattered liberally throughout the UK. The numerous projects presented me with the opportunity of travelling around Britain, acquiring knowledge concerning the various regions and, the icing on the cake, all paid for courtesy of my benevolent employer.

Bernard Evans and I had been assigned the task of performing one of our regular chemical clean up and removal operations at Smith Kline Beecham (SKB), at one of their many research establishments located in Frythe, situated on the outskirts of Welwyn Garden City. This well-known, giant, British pharmaceutical company, possessed wonderful research facilities dotted all over the English countryside, usually lodged in, and based around some ancient country mansion, adorned with wonderfully attractive, ornate, Gothic architecture. The architecture enhanced by the surrounding, spiritually elevating, tranquil scenery.

The site at Frythe was no exception, with its ancient, venerable buildings, accompanied by a wonderful panorama and vista, together with its own luxuriant, verdant forest. The facility was a serene place in which to work and perform microbiological research. A veritable extension to the ancient college spires of Oxford and Cambridge.

The canteen facilities at the SKB sites also tended to

be wonderful places in which to eat. The kitchens providing an extensive selection of amazing culinary delights, affording an eclectic menu, catering to all tastes and needs. SKB certainly knew how to look after its staff. I always looked forward to working at all the SKB facilities, specifically because of their pleasant surroundings and wonderful restaurants, turning up spasmodically throughout the year, to segregate waste chemicals for packaging and transportation. With most of the chemicals destined for incineration at one of our giant furnaces.

This particular week, we found it impossible obtaining rooms for the full duration of the project at our usual hotel, the Bell Inn, located slap bang in the middle of Codicote. Consequently, we were only able to lodge there for the first night of the assignment. I enjoyed staying at the Bell Inn, as it chanced to be located just a few miles up the road from the SKB facility, proving to be an extremely convenient and salubrious place in which to stay. Unfortunately, this precise week, there appeared to be some large seminar or conference taking place, with all of the hotel rooms in the region tending to be at a premium, and extremely difficult to obtain. Our customer contact at SKB provided us with a list of hotels and B&B's in the district.

After phoning many hotels, all of them without success, we finally contacted the Brockett Arms in Ayot St. Lawrence. As luck would have it, the receptionist informed me the hotel had two vacant rooms available. Due to the problems experienced in obtaining accommodation, I reserved the rooms immediately, but not having the faintest inkling for the location of Ayot St. Lawrence or the Brockett Arms. For the first night of our stay at the hotel, we decided to leave work early, in order to search for the location. Our contact at SKB provided us with rudimentary directions to the small hamlet, plus, we possessed a very large, detailed AA map. We are talking pre-advent the technology of sat navs here; a *'must have'* in this day and age. Bernard and I set off confident we would easily locate the small village and pub, anticipating very little difficulty.

Unfortunately, we had not foreseen intervention by the local mischievous kids. Primarily, to while away the boring evening hours, they took great delight in misaligning all the signposts in the area, deliberately pointing them in the wrong direction. Hence, by following the signposts indicating the way to Ayot St. Lawrence, the unsuspecting traveller invariably ended up in the same place, having generally traversed in a complete circle, because of the way the signposts had been aligned, or a more accurate description, misaligned, by the children. We simply could not fathom out, why the roads on the map did not correspond or jive with the directions on the signposts, completely unaware of the nefarious activities perpetrated by the local impish children. After the third time of arriving at the same crossroads, confused and perplexed, with the post indicating the direction we wanted.

Suddenly, I experienced an epiphany, simply turning the post, ninety degrees. The directions then appeared to corroborate exactly with all the nearby hamlets and our AA map. By now, we had travelled miles, wasting almost two hours. It was the latter part of the year, with dusk rapidly approaching, and the sun setting early. We had no wish to be travelling unlit country lanes, looking for an out of the way hotel, late at night. After correctly re-aligning the signposts, we finally made our way towards Ayot St Lawrence and the welcoming sign of the Brockett Arms.

With some relief, we finally reached the end of our magical mystery tour of the Hertfordshire countryside, eventually entering the small hamlet, quickly discovering the Brockett Arms, the only hostelry in the tiny village. It turned out to be a beautiful, quaint, old hotel, many centuries old, exuding wonderful character and charm. I could not even guess when it was built. But it probably existed around or just before the Tudor period. It did, however, possess a modern annex which contained extra rooms, which did detract somewhat, from the character of the main building. Bernard and I were relieved at finally locating the hotel, too tired to consider the arguments of spoiling ancient architecture by erecting modern contemporary buildings, within close proximity to the ancient edifice.

We entered the bar, which also doubled as the reception area, to be greeted by Toby, the Manager, an eccentric middle-aged character who spoke with a plumb in his mouth, and who seemed to be devoid of any common sense whatsoever, and a bit of a hooray Henry. Notwithstanding all his eccentricities and complete lack of organisational acumen, Toby was a genuinely nice person. The bar appeared wonderful, with a convivial, welcoming log fire; a necessary requirement, it being late autumn, with the evenings slowly becoming colder. Substantial, low oak beams supported the ceiling, all enhancing the ambience of the Hotel. The Inn certainly exuded character and charm. Personally, I consider old pubs genial places in which to spend a few evenings. Far friendlier than the modern, artificial travel lodges or large hotels, which appear devoid of atmosphere or charm, I much prefer staying in ye olde country pub.

After signing the register, Toby showed us to our rooms above the bar. The staircase proved to be an old small, winding timber, spiral stairway. The bedroom rooms also had low ancient oak beams. The doors, instead of locks, had latches, all these accoutrements helping to enhance the rich atmosphere emanating from the ancient, atmospheric building.

That evening, we had a few drinks in the bar, followed by a very nice meal, then retired early. It had been a long day, both Bernard and I were exhausted. The next morning, we finally had breakfast at eight thirty, having requested it for seven thirty. But then, that was Toby for you.

We had almost completed our work on that third day of the project. It would only take a few more hours on the fourth day, and our work would be finished. After which, during the afternoon, we could drive back to South Wales at a leisurely pace, following an early lunch in the SKB canteen. Having worked hard on that third day, with everything nearly completed, Bernard and I decided to leave early at four o'clock. This time, we knew our way to the hotel, so it did not take us as long to get there. Upon entering the bar, prior to going to our rooms for a rest, Bernard suggested we have a drink. At the bar, we came across one solitary customer, comfortably perched on a stool. The customer appeared to be completely relaxed, and well at home. Bernard and I stood at the bar, with no sign of any service. We talked as loudly as possible, endeavouring to obtain some attention from any hotel staff who happened to be in the vicinity.

'You'll have a long wait. Toby is not around.' explained the lone customer, while exuding an ingratiating, friendly smile. His teeth dispersed sparingly throughout his mouth, resembling the headstones in an old, dilapidated cemetery, that is to say, deeply ingrained, dotted here and there, with large voids in between.

'What would you like to drink?' he enquired nonchalantly, exhibiting a broad country accent. Bernard and I looked at each other slightly bemused, not knowing whether to take him up on his kind offer.

'A Pint of Cider would be nice,' replied my

colleague, quickly deciding upon his ultimate course of action.

Following Bernard's lead, I scrutinized the many pump handles located on top of the bar.

'I'll have a pint of King's Raven,' I answered. Having consumed a variety of beers the previous first evening, I had decided upon that brew as being the best, due to it exuding a wonderful oaky flavour, and much to my liking.

Our reserve barman went behind the bar, pulled the drinks we requested, then passed them over the bar to us before quietly returning to his perch. We told him to put the drinks on a tab, but he replied with his cordial smile,

'If Toby's not here, that's his lookout.'

We began chatting to him, and the customer explained he had been at the bar since two thirty and had not seen hide no hair of Toby in all that time. It was just after four thirty, so the bar had been unattended for well over two hours.

The regular explained Toby's father-in-law owned the hotel, which he let Toby manage. The father-in-law lived on the south coast of England and had probably given Toby the task of running the pub to keep him out of his way. Although, the drinker believed Toby was not an ideal businessman, and probably running the pub at a loss. He could not comprehend why Toby's father-in-law continued to employ him as the manager. I also wondered how Toby ran the place with his regular customers taking drinks, without paying for them. We had our drinks, chatting to the regular, who mostly discussed Toby and his idiosyncrasies. We both finally retired to our respective rooms, before the evening meal.

Bernard's room was next door to mine. After agreeing to meet him in the bar at seven thirty, for a few more drinks, before taking our evening meal, I was about to enter my room. Bernard opened his door, which only had a latch, no key, only to observe a man in his late twenties or possibly early thirties, sitting on the bed, quietly reading a newspaper. I could see Bernard displaying a bemused expression, then taking a step back, to reconsider the room number on the front of the door.

'I thought this was my room?' he then stated, in a slightly perplexed tone, to the interloper.

'Sorry,' replied the squatter, his reply, bordering on an apology, 'But this is my room.'

Bernard and I went downstairs, to ascertain what was going on, and hopefully discover the whereabouts of his missing belongings. Upon reaching the bar, we discovered Toby had miraculously appeared. Bernard tried explaining, as calmly as he could, concerning the circumstances of his room being purloined by a stranger, and asking the whereabouts of his personal belongings.

'There's been a bit of a cock-up. Your things have been put into another room,' replied an embarrassed, apologetic Toby, before presenting Bernard with the key to his replacement room, situated in the new annex.

I could not understand why the new customer had not been allocated Bernard's new room, without Bernard having to relocate to another room. Why hadn't the new customer simply been given the room in the annex instead? Then again, that tended to be Toby, with his anarchic style of running things.

Once all that had been resolved, I returned to my room, for a short siesta before showering, changing and going down for the evening meal.

At seven thirty, I went downstairs. At the bar, I ordered yet another pint of the wonderful King's Raven beer, and awaited Bernard's arrival. He appeared a few minutes after me. We had a couple of drinks before ordering a meal.

The meal arrived, and we began consuming it as if it were our last, for Bernard and I were ravenous following our days' work at SKB. While sitting at the large table, a man in his late twenties approached us, enquiring if he could sit at the same table as us, explaining all other seats had been taken in the restaurant. Now, unbeknownst to me, this was the same person, who had purloined Bernard's room. I later found out, Bernard recognized him, but the man however, for some unknown reason, appeared not to recollect Bernard.

We began chatting. It turned out he was called Steve, and the Sales Director for a company which supplied Process Control Equipment, such as flow meters, temperature gauges, pressure gauges, control valves, regulators etc. He too, had been unable to get a hotel room in Welwyn, before, like us, finally discovering the Brockett Arms. When he found out we worked for a chemical company possessing a large amount of chemical engineering equipment, he became extremely interested. Once a Salesman, always a Salesman. Bernard and I were able to supply him, give him contact names for the engineers at the South Wales facility and the other incineration facility near Southampton. The young man began quizzing us about Chemkiln, and its parent company. Being an out and out salesman, he felt unable to miss an opportunity to sell his wares and obtain information about a potential lucrative customer.

The evening progressed convivially. Steve bought us a drink, we bought him a drink, he bought us yet another drink. So, the evening progressed, with all three of us eventually consuming a considerable amount of alcohol.

Now, the previous evening, I had been informed by one of the locals about a ghost who allegedly haunted the old building. Purportedly, the ghost was thought to be that that of a devout Catholic monk who had been hung in the hotel during the Reformation, primarily because of his unwillingness to denounce and renounce his Roman Catholic faith. It was rumoured the spectre of the monk often appeared in one of the bedrooms located directly above where he had been executed. Towards the end of the evening I casually slurred out the story to our new-found drinking companion, believing it to be a good topic of conversation. Now Steve could not be described as a wimp, well over six feet tall, built like the proverbial brick shithouse and in his late twenties or early thirties. He suddenly turned pale, and not from the amount of alcohol he had consumed. He then began questioning me nervously about the story and asking others in the bar if the story was true. Bernard returned from the toilet and asked what was going on. I told him I had informed Steve about the story concerning the apparition of the executed monk. This was the result; a gibbering idiot.

'Which bedroom does he appear in?' Steve asked the barmaid, his demeanour now suddenly bordering on hysteria. The barmaid informed him she did not know.

Bernard, a person renowned within Chemkiln for being a practical joker, knew which room Steve was staying in, after all it had been his room the previous evening. 'Number three.' he casually remarked, without a hint of humour on his face. That was it, the comment pushed Steve completely over the edge. Although, I must point out, Steve needed very little pushing.

'I want to see the manager. I want to be moved immediately to another room in the annex!!' he shouted so loudly, everyone in the pub could hear him. Eventually Toby finally appeared to find out what was going on. Steve explained to him his fear of the supernatural and the ghostly apparition of the monk, informing him he wanted to be moved to another room in the new annex, also informing Toby he wanted the hotel staff to do all the packing of his personal belongings. Steve absolutely refused to go back to his allegedly haunted room where, he believed, the unwanted spectre of the dead monk lay in waiting.

Toby agreed to his customer's demands. While Toby packed his belongings, Steve remained at the bar talking to the locals, which is the worst thing he could have possibly done, for they did nothing to relieve his obvious fears. One enthusiastic storyteller even went as far as showing him the hook from which the poor unfortunate monk was supposed to have been hung all those centuries earlier. To my untrained and somewhat intoxicated eye, the hook looked like a small nail which may well have been bought from a B & Q or the local hardware store, just a few years earlier, and not four centuries before that. Poor Steve was in such a state and believed it all, taking it all in, while the locals casually related more stories concerning the unfortunate monk. Unfortunately, with each telling, the stories became more outlandish, bizarre and unfortunately, from Steve's point of view, more blood-curdling. When Toby finally appeared with his belongings Steve nervously followed him to his new room in the modern annex. That was the last time Bernard and I ever set eyes on poor Steve.

The next morning at breakfast, exhibiting a slight hangover Bernard and I waited for Toby to appear to provide us with breakfast. As usual he appeared at five to eight, when we had requested breakfast for seven thirty. By the time he eventually commenced cooking we finally began eating at eight twenty, much later than we would have liked.

We began talking to him about the monk and asking him how the story of the ghost came about. It seems there had been four sightings on four separate occasions by different, guests. One guest even stayed awake all night with his bedroom light on and had dark marks under his eyes the next morning after seeing the apparition of a hooded monk walking, or rather, gliding through his room. Toby stated he did not know whether the ghost really did appear but the guests all looked terrible after their ghostly experience. After consuming our breakfast, I asked Toby for the bill explaining we also had a considerable drinks tab together with meals, asking him if he could somehow put meals and drinks together disguising how much alcohol we had consumed. To which he agreed. Watching Toby going through all the slips was a remarkable sight; and going something like this.

'Pint of this plus pint of that that's approximately four pounds plus meal that's about fifteen…' To Toby everything carried the tag 'approximately.' Finally, he did an approximate total. Whether it worked out to his benefit or Chemkiln's, I have no idea, but I just wanted us to be on our way back to South Wales as soon as possible.

Before we left, I asked Toby whether it had in fact been Steve's room, which the Monk was supposed to haunt. Toby thought for a moment trying to recall the occupants who had the experience of sighting the monk.

'As I recall,' continued Toby, in his rather posh hooray Henry accent, 'I don't think it was his room. I believe it was room number five.'

At that moment, it was my turn to go a whiter shade of pale. Number five had been the room, in which I had slept the two previous evenings!

Alas, I have never had cause to stay at the Brockett Arms since. I would dearly love to go back there, for the place, exuded such character and charm, unlike many modern hotels which have no such appeal being sterile places, and disappointingly, without their own resident ghost or entity.

CHAPTER 13

During my early days working for Chemkiln, it soon became evident Dr. Phipps appeared to be loathed intensely by his subordinates. In addition to not being liked one iota by most people within the company, Damien proved to be an arrogant individual. A martinet, with a huge ego. A control freak, extremely ambitious, with aspirations of one day becoming Managing Director of Chemkiln, even possibly becoming MD of the main holding company. Damien's aspirations seemed infinite and the *sky's the limit,* his career ambition recognising no bounds. Unfortunately, because of these desires, Damien felt almost compelled to trample on anyone and do almost anything to attain his goal of reaching the upper echelons of the management tree. He exhibited traits of psychopathic tendencies, which it seems to be the norm with high flyers. Apparently, almost fifteen percent of company directors and CEO'S have such tendencies. A psychopath does not necessarily indicate a murderer. One dictionary definition of a psychopath; *deems he or she to be a person characterized by persistent antisocial behaviour, impaired empathy and remorse, possessing bold, disinhibited, and egotistical traits.*

The General Services department turned out to be an extremely interesting, entertaining and fertile place in which to work; I thoroughly enjoyed observing all of the office politics, ambitions and intrigue. As the months and years went by, those intrigues eventually manifested themselves fully, causing massive upheaval and turmoil within the department.

As mentioned in an earlier chapter, within a few weeks of joining the General Services Team, I attended a Team Building course run by the Industrial Society, together with managers and members of other departments in the company. It became evident from the course that quite a few of the managers had acute problems with their management skills; amongst them Dr. Phipps. The upshot being, the company forced Damien to attend additional courses provided by the Industrial Society because of his attitude and general behaviour towards his subordinates and others within the company, an evident shortfall in his man management skills. After completing the team building course and being warned about his over-assertive, bullying behaviour by higher management, Damien tried desperately to be as nice to people as he possibly could, especially towards his immediate staff. Unfortunately for Damien, these remedial actions proved far too late, as Rupert had already put in a few complaints about his attitude and the way Damien frequently patronized, berated, generally humiliated and bullied people in front of their colleagues. The seeds had been gradually sown for his slow demise.

Because Damien's ambitions knew no bounds, almost two years earlier he had overthrown his, then immediate boss, Jim Beam. The name, Jim Beam, given to Damien's boss by his parents, and the same name also given to a bourbon whisky, ended up being prescient and prophetic. For in later life, Jim ended up with a drink problem, similar to many people in the waste industry, due principally to the nature of the business. Keeping customers happy and entertained usually meant consuming large amounts of alcohol, while attempting to be sociable in the attempt to acquire lucrative contracts. After years of socialising and heavy drinking, Jim soon experienced great difficulty in giving up the demon drink. His downfall came about whilst supervising a particular project on Deeside working for a giant chemical company. Damien had initiated Jim Beam's ruin by informing the Project Manager of the chemical company about Jim's drinking problem.

One morning, Jim turned up on site after having taken a couple of drinks from his hip flask, which he kept on his person. Before commencing work, Jim often found it necessary to drink some alcohol. The supervising Project Manager for the site, having been made fully aware of Jim's problem by Damien, detected the drink on his breath, immediately sending him off site. The client's Project Manager reported the incident to Chemkiln and Jim subsequently lost his job. In those days, companies tended to exhibit little understanding or care about their personnel and any difficulties experienced concerning alcohol; offering no counselling or aid. Because of Jim's dismissal, Damien gained immediate promotion, jumping into his bosses' vacant shoes; his ultimate intention from the outset.

When I joined the company, Damien had moved on with his aspirations, his eyes now firmly set on Nigel Davidson's position, that of Sales Director. The Sales Department encompassed the General Services department and Damien reported directly to Nigel. The title of Manager proved insufficient a label and not grandiose enough for Damien. He had his eyes resolutely fixed on becoming a director, with a seat on the board; putting it simply, he coveted Nigel's job.

Damien was always looking for opportunities to get promotion and whenever an opportunity presented itself, he would invariably jump at it. Before I proceed any further with my story, we must flash back earlier in time to when Jim Beam ran the General Services Department. At that time, there existed a long-running personal feud between Marilyn, one of the other supervisors, and Damien. The intense hatred between these two, highly ambitious individuals, being reciprocated in equal measure. When Damien finally ousted Jim Beam, eventually becoming head of General Services, the advancement in Damien's fortune and career proved too much for Marilyn. Within a short period of time, she jumped ship, changing departments, transferring to the Sales Department, finding herself unable to work under Damien, a man she loathed with a vengeance. Damien exacerbated the situation, speeding Marilyn's departure by performing his usual trick of constantly humiliating and berating her in front of her colleagues. It must be pointed out Marilyn was no angel by a long shot, with ambitions almost equal to Damien's. Undoubtedly, had the roles been reversed, she would have probably done precisely the same to him. These were two huge egos, continually at loggerheads.

So, Marilyn joined the Sales Department as a Technical Sales Representative but still took every opportunity to stick the career killing knife into Damien whenever an opening presented itself. Damien did likewise with Marilyn. Evidently after the Team Building course, the fact Damien had been reported about his general man management skills by the Industrial Society lecturers, gave Marilyn immense gratification and joy.

Within a year or so of me joining Chemkiln, rumours began circulating about Nigel Davidson and Marilyn, with the two of them frequently observed socialising and drinking together in Cardiff or other locations. The rumours persisted, in truth, mainly generated by Rupert St. John Smythe who lived in Cardiff and often witnessed the two of them out together. It was difficult to prove there was actually anything going on between Marilyn and Nigel as these sightings usually occurred during sales conferences at the incineration facility, which necessitated Nigel staying in one of the local hotels, usually Cardiff, the capital of Wales. Living as he did north of Birmingham.

Nigel was in his early-forties, married with three

children. While Marilyn was in her mid-thirties and single, but certainly not inexperienced or reserved when it came to members of the opposite sex. As Marilyn lived in Cardiff, it could be all innocent, and she went out with Nigel simply to keep him company and ingratiate herself for promotion or a pay rise; being an opportunist, much like Damien.

I suspect, actually, more than suspect, I know, Rupert continually fuelled the rumours, concerning Nigel and Marilyn, more to wind up and infuriate Damien, for no other reason than to activate and trigger his boss' intense paranoia. For like all control freaks, Damien wanted to know what was going on and be in complete control. Knowledge is power, as the saying goes. The fact, a woman he hated intensely and with passion and who reciprocated those feelings could possibly or more likely, be having carnal knowledge with his immediate boss caused him great worry and distress. Undoubtedly, in Damien's paranoid mind, Marilyn was poisoning Nigel's mind against him, most probably during the pillow talk.

Although I suspect Damien was probably correct in his supposition, having talked to Marilyn, she would have taken any opportunity to slander Damien. She certainly had a lot of information about him with some intimate knowledge concerning some of the skeletons which lay dormant in Damien's personal cupboard. The circumstances played constantly on his mind. But it was all conjecture, innuendo and supposition. Was there anything between Nigel and Marilyn? Nobody within Chemkiln really knew or had any real tangible proof. Every time Rupert came across Marilyn and Nigel out on the town together, he made a point of making Damien aware of the fact, either by talking loudly in the office or telling Damien directly to his face. Gently stoking and fermenting the fire. As the weeks progressed Damien's features began to alter, and he began taking on the appearance of a troubled, haunted man. Even if they weren't actually sleeping together, what was Marilyn divulging to Nigel about him during their drinking bouts?

An incident transpired which eventually resulted in Damien leaving Chemkiln. Damien had travelled to Thailand in an attempt to win a lucrative project for the removal from that country of transformers containing PCB. In addition, it turned out to be a bit of a jolly, with Damien attending the rugby sevens taking place in Hong Kong while en route. While Damien jet-setted around the globe, Matt Kent, one of the department's supervisors and also a salesman for the department, had gone to Nigel's office.

That day, the Director was on site and Matt needed some advice, so went to Nigel's office to discuss the tricky issue with him. Matt knocked on the closed door then entered the room, only to discover Nigel and Marilyn in a full, highly intimate embrace, kissing quite passionately but not full coitus. There could now be no doubt, they were more than just boss and subordinate or even simply good friends, they were evidently indulging in a full-blown, gold-plated affair. The two lovers, after briefly catching sight of Mark, quickly broke away from the clinch, moving rapidly away from each other as people in those circumstance tend to do, hoping, by performing such quick actions, their indiscretion would remain unobserved. Quite risible really. Matt immediately vacated the room, highly embarrassed, not knowing what to say or do. After discovery of the office intrigue and scandal involving the Sales Director and his female subordinate, Matt did a remarkable thing, which, to this day, I cannot comprehend. He sent an e-mail to Damien describing the day's events. Why he did this, I do not discern, as he too disliked Damien. Whether he did it to wind him up just as Rupert had done, I do not know? I am being underhanded and cynical when I say Matt could have kept quiet and used this knowledge to his own advantage possibly furthering his career. Skeletons in cupboard, and all that. Instead, he informed Damien.

Immediately, Damien caught the first available flight back from Hong Kong, where he had been enjoying himself, watching the rugby sevens. The opportunity to possibly oust Nigel and hopefully take his job proving too good a chance to miss. As Damien frequently liked to say, '*carpe diem.*' (seize the day). The first thing he did on getting back in the UK was to arrange a meeting with James White, the Managing Director of Chemkiln and who normally based himself at the other incineration facility located on the south coast of England.

Apparently, so Damien kept trying to persuade people. He thought it morally disgusting and reprehensible that a married Director of the company with two young children had been indulging in an extra marital affair with one of the female members of the sales teams. Not that Damien exhibited any religious inclinations or indeed any sort of moral indignation or outrage concerning the situation. In truth, he too had often indulged in extra marital activities whenever the opportunity presented itself. No, Damien's main concern related to his nemesis and arch enemy Marilyn, who was evidently sleeping with his boss, and God knows what she was telling him. His paranoia leapt to the surface. Damien would use any method at his disposal to try and put a stop to this unwanted liaison between Marilyn and Nigel, and hopefully, in doing so, take another step up the career ladder achieving further promotion, by getting Nigel dismissed.

Unfortunately for Damien, James White and Nigel Davidson, as well as both being Directors of Chemkiln, had a close personal friendship. In fact, it had been James who enticed Nigel to join Chemkiln in the first place, the two of them previously having worked together at BP. James was not full of self-righteous disgust with Nigel's behaviour as Damien had hoped. Far from it, in fact, helping protect his fellow director and friend. I suspect James knew about the affair long before Damien. Now both Directors were gunning for both Damien and Matt, the latter being the instigator of all the upheaval and turmoil within the department.

A few weeks after Damien had almost demanded James White remove Nigel Davidson, the MD Suddenly created a new position between Damien and Nigel. Giving the new position to one of the sales people, a guy named Sid Nightingale: another person in the company who loathed Damien. Damien's plan of ousting Nigel had grossly backfired,

Sid Nightingale had been in the Sales department bringing in copious amounts of bulk waste for disposal by the company. A hot shot salesman. He had, in all fairness brought in a large amount of business for the company and had told Norman he could get even more business for the specialised General Services department. Waste PCB from transformers was in a steep decline since the heydays of the eighties and new waste business needed to be brought in for the specialist department of General Services whose main forte involved specialized hazardous waste projects. At first, Sid's tenure began tentatively. Damien, because of his vast knowledge of the department, felt confident enough to bamboozle and flummox his new boss. I personally thought he did not show the respect he should have towards Sid, constantly trying to humiliate and belittle him whenever possible.

As the months progressed, Sid gradually began to get his feet under the table becoming more au fait with the running of the department. As he became more acquainted with the department and more confident, he began exerting his authority within the department and started overriding lot of Damien's ill-advised decisions which Damien proposed with the intention of bringing the department into disrepute and hopefully restore him to his rightful position, back in charge. Which is where he believed unreservedly he should be.

Sid began to become more aggressive and assertive towards Damien, knowing full well the mind games Damien continually indulged in. After about a year or so, Sid eventually prevailed, resulted in Damien finally handing in his resignation and getting a job with the United Nations, and a department involved with the removal of toxic pesticides from underdeveloped, third world countries. Damien always dreamed of holding a UN passport, so he was no too upset, even though it meant spending a large amount of time away from home. Damien's wife Debora was also happy as she did not have to see too much of her husband with him being based mostly abroad. A Godsend as far as she was concerned.

Matt Kent, also, did not last long with the company afterwards. He too had difficulty with James White and Nigel Davidson. Although it was Matt's wife who eventually brought about his demise with Chemkiln. During this time, Matt and his wife had separated and lived apart. One night, Matt's wife, Anne suggested they meet up for a meal to try and resolve their marriage difficulties. The meeting ended up badly and acrimoniously, with Matt angrily storming out and driving home and but only after having consumed a few beers. Hell, hath no fury like a woman; well you know how it goes. Anne immediately phoned the police informing them about a suspected drunk driver, providing them Matt's car registration number. Subsequently, he got stopped and discovered to be over the drink, drive limit. Eventually losing his driving licence. His job depended on him being able to drive around the country and hence the instant he lost his driving licence, he also lost his job with Chemkiln. Neither James White or Nigel Davidson felt inclined to help him out after the problems he had stirred up by telling Damien about the affair.

Personally, I considered Sid to be a reasonable boss and we got on well together, coupled with the fact, he always treated me fairly, which was not always the case with Damien. Although, I was not alone in that respect, Damien treated everyone unfairly, unless they had something on him. Unfortunately, eventually, Sid also fell foul of the top management and within a short time, he too departed the company after failing to fulfil his overambitious promises of vastly increasing business for the General Services Department. Promises which never came to fruition. Because his promises never came to be fulfilled, Chemkiln eventually removed Sid from his job, forcing him to leave the company.

Ironically, within a few months of Damien quitting Chemkiln, Nigel Davidson died of cancer and Chemkiln employed another person from outside the company to fill his position. So, Damien should have sat it out and waited. He may well have possibly been elevated to the coveted position of Sales Director. But, we will never know what may have indeed transpired. The proverbial, 'what if,' scenario.

Don't you just love office politics and intrigue. I certainly do?

CHAPTER 14

There is a wise old saying, and it is a philosophy with which I agree implicitly. It goes something like this, 'Be careful how you treat people on the way up, for you could possibly meet them again on your way back down.'

I would also put in my own personal supplement.

'Be careful how you treat people generally on your journey throughout life, for they may well overtake you further along the route.'

The ramifications of not adhering to these simple, home spun, philosophical judgments concerning one's attitude towards life have revealed themselves to me with significant effect, time and time again, mostly through observing other people and their mistakes in not following these simple principles. One such person who fell afoul of this law turned out to be Albert, but then again, who else could it possibly be?

One Friday lunchtime, Ioan, Albert and I were approaching the main entrance to the offices at the facility. We had all been working on site, having completed the week's projects away from the site on Thursday. The three of us noticed Bernard chatting to one of the senior sales guys, a certain Sid Nightingale in front of the latter's company car. Bernard beckoned the three of us towards him

'Sid's got all these display stands which he needs to put back in the sales office. I said we would give him a hand.'

We all peered into the back of the large estate car, and sure enough, the vehicle appeared to be jam packed with some impressive and not insubstantial display stands which had been required for an industrial sales exhibition held that week somewhere in the Midlands.

We all concurred, he did indeed have a large amount of heavy equipment to transfer and we all agreed to help. I make the remark, 'all of us,' which is not correct, for there was one exception, Albert, not only did he refuse to help, but was quite derogatory, vitriolic and insulting with his refusal.

'He can fuck off, he's only a jumped-up salesman. I'm not helping him!' Having made that derogatory, unambiguous statement, Albert sauntered off on his own merry way. Albert was your stereotypical sycophant, always crawling and sucking up to his superiors but dismissive towards individuals who appeared to be of no benefit whatsoever to him. At that time, as far as Albert was concerned, the poor salesman, most definitely, fell into the latter category. I emphasize, at that time, for within a matter of two months, Sid received promotion, elevated to becoming head of our department, with even Dr Phipps reporting to him.

Albert's face was an absolute picture when informed Sid Nightingale would be our new head of department. His jaw dropped, and he looked as if he would suddenly burst into tears.

From the instant Sid took on the mantle of department head, Albert somehow managed to orchestrate having very little contact with him, always staying well clear of his new boss.

Poor Sid, I did sympathise with him. His main forte was that of a salesman, with his man management skills bordering on being virtually non-existent. He failed miserably when it came to gain any respect from his subordinates, particularly Damien, who despised Sid immensely particularly after being deposed so ignominiously by him. As pointed out in a previous chapter, from the outset, Damien took every opportunity to belittle and demean his new boss, fully aware of his new boss' lack of experience with the General Services' side of the business.

I was put in charge of a chemical decontamination

project for BP located near Sunbury-on-Thames, on the periphery of London. The facility being one of the sites which dealt with oil and exploration and extraction. There was a considerable amount of obsolete oil samples from oil rig production platforms situated around the globe. The facility at Sunbury required extra storage and the management considered it time to extirpate the site of the ancient samples acquired throughout the previous decades of the twentieth century, and destroy them by elevated temperature incineration, which is where Chemkiln came into the equation.

At that time, the General Services department had a lot on its plate with full order books and limited amount of man power to carry out the promised jobs. Conversely the other incineration facility belonging to the company located on the south coast had a lull in its capacity leaving some of its labour force with very little work to do. After being approached by the General Services department, the incineration facility sent three operators to assist me, Derek, Mike and Tony. They were good guys, all ex-army and most amenable in temperament, performing tasks requested of them without any form of rancour or argument whatsoever, and occasionally coming up with helpful suggestions. During his brief period in the army, Mike had been near an explosion impairing some of his hearing in his left ear. Consequently, the others relentlessly took the piss out of him, in a form of army banter. In return, Mike gave as well as he got. Your normal armed forces sense of humour and banter.

During the couple of weeks, the project progressed well, and I became friendly with the team. We all stayed in the same Hotel, located on the outskirts of Esher, often going there for a meal, spending most evenings together.

Sid had only been in his new position and my boss

for a few weeks, and I hardly knew him. One day he phoned enquiring how the project was progressing. We had a reasonable chat about the job and how long I expected it to take. He also asked how the guys from the other facility were performing and if they were all behaving themselves. He then casually asked me where we intended going that evening. I informed him we intended going to the Bear Hotel for a couple of drinks before heading for one of the nearby Indian restaurants for a curry.

He completely took the wind out of my sails by informing me he would be travelling home that afternoon and that he lived near Esher and would more than likely join us for a drink that evening. I was a bit apprehensive as I did not know Sid at all and had no idea of his temperament and how to behave towards him. A situation all working people must go through sometime or other during their careers. We agreed to meet him in the Bear about seven thirty that evening. I had to inform my team about the rendezvous and pleaded with them to be on their best behaviour as I had absolutely no idea what Sid was like. I virtually pleaded with them not to embarrass me or land me in hot water with my new boss, fully explaining the situation to them.

Derek casually remarked he would not be joining us that evening and unashamedly added he had a date with a married woman from Portsmouth with whom he had been having an affair for some time. This project gave them the opportunity of spending the night together paid for courtesy of Chemkiln.

The others obviously knew about the affair and took it in their stride. Derek and Tricia, the woman in question were both married, and a chance of spending an illicit night of sexual passion together was too good an opportunity to be missed.

Tony, Mike and I arrived at the Hotel at about seven

that evening. We had only been there for about ten minutes when Derek suddenly and unexpectedly turned up with his mistress. I must confess Tricia was a real stunner. I guessed her to be in her late thirties, with shoulder length, blonde hair, neatly and expensively coiffured. Dressed immaculately, she exuded some style and Derek was obviously keen to show her off. 'Look at the type of woman I can snare,' sort of attitude and gave the air of the cat who had pinched the cream.

Despite her obvious class and projecting a demure countenance, Tricia proved to be quite chatty and not standoffish, exuding charm and quiet confidence, obviously comfortable in her own skin. She was gorgeous and knew it, but not aloof, unlike some attractive woman who often look as if they have a bad smelling kipper under their nose.

We were all having a good chat, meanwhile I kept reminding the guys to be on their best behaviour in front of my new boss. At precisely seven thirty, as arranged, Sid entered the bar of the hotel. As soon as he came in and I made the introductions and, being a good boss, he immediately bought a round of drinks.

Sid and I talked for a brief time, mostly discussing the project and the guys. I gave some brief descriptions and insights into my colleagues. It was during our talk, Sid, referred to Tricia and Derek.

'Don't tell me he's pulled already since he's been here?' He remarked.

'You don't want to know, Sid, it's a bit more complicated than that. They have known each other for some time,' I explained quickly.

'Are they an item?' he pressed further.

'Sort of, they are both married, but not to each other.' I added, considering it wise keeping him in the loop.

After my statement, a wry grin appeared on Sid's face.

'I see it's like that is it, they fancy a comfortable

hotel room at Chemkiln's expense?'

'Something like that,' I added, wishing he would change the subject, uncertain how he considered the romantic, although immoral, situation. Perhaps Sid was highly religious, considering the situation highly sinful, I pondered.

By this time, I had drunk a large amount of my pint, most probably through nerves. Sid looked at my virtually empty glass.

'Fancy another drink?' he enquired.

'Yeah sure, the same again please,' I replied.

Sid told the others to drink up and made his way to the bar to get the round in.

After all the drinks had been sorted and distributed, Derek went to the toilet. It was at this point that Sid casually sauntered over to Tricia, and began indulging in conversation with her. At that time, I was well out of earshot and talking to Tony. I surreptitiously glanced over to my new boss and the voluptuous Tricia. The two of them appeared to be getting along like a house on fire.

Suddenly and unexpectedly, Tricia's demeanour altered dramatically, her voice increased in volume and octaves, enabling it to be heard above the general hubbub of the room.

'You can fuck off!!' she shrilled out, her voice raising an octave or two higher.

A silence coated with huge embarrassment quickly engulfed and overwhelmed the room. With that, Sid quickly made his way back towards me, his face, an extremely bright red, I surmised, not because of the alcohol, but because of what had just transpired.

I tried desperately to act nonchalant, meanwhile, the general cacophony of noise, slowly returned to the room, with the customers in the room, periodically casting furtive glances in Sid's direction.

Poor Sid, it seemed to be his lot in life with

individuals forever advising him to 'fuck off!!' or 'go forth and multiply.' Both Sid and I consumed our drinks in a distinctly embarrassed silence. Meanwhile Derek reappeared from his relief break and from his malevolent glances directed in Sid's direction, he had been made fully aware of the situation by his mistress.

From that instant, Sid ingested his pint in record time.

'Well I think, I'll skip the meal and head home,' he informed me trying to maintain an air of nonchalance, before making a speedy exit and returning to the relative safety of his domestic abode.

I never did discover what transpired between Sid and Tricia. Both Derek and Sid kept tight lipped about the whole situation. But I suspect Sid tried it on with Tricia, and in return, experienced short shrift.

Within the week, the project at BP Sunbury reached completion, and we all returned home and that was the last time I ever saw of the ex-army personnel from down south.

CHAPTER 15

I discovered first hand, while employed by Chemkiln, the unhealthy reliance and over dependency which the industrialised and developing nations of the world place upon oil. The dire consequences when the earth's limited oil reserves finally become exhausted, and how human beings should look towards alternative, renewable, green sources of energy to survive.

During most of the twentieth century, many major confrontations tended to erupt mainly because of oil. Hitler with his dogmatic, detestable, political doctrine and intense hatred towards Communism, embarked upon operation Barbarossa, the planned invasion of Russia, primarily to capture the immense oil fields located in the Caucuses and Baku, although, the possible overthrow and elimination of another despot sustaining a political dogma which he detested manifested itself as an extra bonus. Additionally, Japan as part of its overall strategy carried out an unprovoked attack on the American fleet at Pearl Harbour to obtain free reign over the oil fields in Borneo and Sumatra, consequently with the prospect of achieving unhindered, cheap supplies of the black gold to their homeland.

Even the first Iraqi War in the nineties, allegedly to remove Saddam Hussein and the Iraqi army from Kuwait, arose in reality, because of the vast amount of oil reserves lying beneath the ground in that region. In the new millennium, wars are still being waged mainly because of this literal thirst for oil and the energy it generates, subsequently propelling the industrialised nations and their economies along.

In the autumn of 2000, General Services had its regular eight weekly agreements for the removal of toxic chemicals from the Astra Zeneca, an extremely large pharmaceutical research facility which employed approximately five thousand staff and located near Macclesfield. When I say Macclesfield, in truth, the site could be designated as being geographically located much nearer Alderley Edge than Macclesfield. Alderley Edge being the exclusive, select, expensive suburb of Manchester and where many top league football players lived and still do to this day. At one time, even David Beckham and his perpetually smiling wife Victoria (I jest), resided there.

At the same time, during that autumn of 2000, the price of fuel began spiralling out of control, reaching exorbitant amounts. Farmers, hauliers and anyone associated with the transportation business began seeing their relatively small profit margins diminishing rapidly. This reduction in their living standards generated immense feelings of resentment and antipathy towards the government. The extremely high taxes being levied by the Inland Revenue exacerbating the high price of fuel.

At the beginning of the year 2000, tax accounted for 81.5% of the total cost of unleaded petrol. Things came to a head on Friday eighth of September when farmers and hauliers began blockading the Stanlow Refinery near Ellesmere Port. The blockade escalated throughout that weekend to more refineries and other fuel storage facilities located throughout the UK. By Monday eleventh of September 2000, nearly all garages throughout the UK began running short of both petrol and diesel, with most outlets forced to implement some form of basic rationing.

Bernard Evans and I had been assigned the task of carrying out the project at the Astra Zeneca facility on the Tuesday. Because of the fuel situation in the UK and the uncertainty, we decided to travel to Manchester that Monday afternoon. Bernard drove the Renault Turbo diesel van, vital for transporting all the required equipment for packaging hazardous chemicals, while I drove my Diesel-powered Rover 75. Unfortunately, we could only manage to find accommodation in the town of Congleton, and a fair distance from the pharmaceutical company.

Normally, this would not be a problem, but with petrol and diesel in short supply, we needed to conserve as much fuel as we possibly could in order to make our way back home to South Wales once the project had been completed. There appeared to be no end to the blockade in sight with the negotiations between the Labour government and the protestors seemingly to be at an impasse. I decided we should leave the Renault van parked on the Astra site, and use the Rover with its higher miles per gallon for the daily commute between the hotel and the facility.

On my way up north, I had somehow fortuitously managed to fill up with diesel at the Hilton Park Services, one of the M5 service stations near Birmingham, and one of the last service stations to have reserves of petrol and diesel left in their underground tanks. By the time I arrived at the hotel in Congleton, I still had almost a full tank of diesel. Bernard also managed to obtain diesel for the company van and, so he too had a reasonably full tank of fuel.

The job went to plan, without any mishaps or unforeseen problems, and by the Thursday morning, with the task completed we headed back to South Wales. The dispute and consequently the embargo ended on that Thursday morning, with the government allocating police escorts to tankers and agreeing to review the burdensome taxation being imposed upon petrol and diesel which they reduced in the November budget. However, at that time virtually every garage in the country had completely run out of both diesel and petrol and it would take days or, in some cases, weeks for the garages to replenish their stocks to where they were before the dispute began.

We completed the job about eleven in the morning. After getting all the legal, regulatory, hazardous paperwork signed by our customer contact, Nigel, we headed south in our respective vehicles. The assumption being there would be no fuel available to buy during the journey. A correct assumption as it transpired. I had split up with Stella and now lived with Bridget and on the outskirts of Llanelli, a lot further west than where I had previously lived with my former partner. It was undeniably going to be a nerve-wracking journey back home to Burry Port.

I let Bernard drive off first, following him, just in case he experienced any difficulties with his fuel consumption. Whilst driving through the outskirts of Congleton, I noticed numerous City Link vehicles with their distinctive green and yellow livery queuing at one of the garages which had obviously just received, or about to receive a delivery of fuel. Further along the route, after joining the M6 and driving past the Sandbach Services, I noticed many vehicles parked on the garage forecourt. Upon closer inspection, it became evident they had all been abandoned there after running short of fuel with none available in the services.

Bernard had left some time before me, and now driving some miles ahead of me on the motorway, a motorway completely devoid of vehicles. Throughout the years, I frequently travelled on the M6 and on all those previous occasions, whatever the time of day the motorway tended to be completely congested with miscellaneous vehicles of all varieties. This was chillingly eerie and totally surreal. It was noon on a normal workday, a Thursday, with the occasional sighting of a car travelling at a sedate, reasonable speed to reduce their fuel consumption. Both carriageways, north and south bound, to all intents and purposes, were completely empty. It was as if Armageddon had finally arrived and as if some cataclysmic event or horrific plague had decimated the population of the UK. The most frustrating part of the situation was I too was forced to travel at a sedate speed to conserve my fuel, travelling at a measly fifty-five mph on a three-lane, empty, deserted motorway, completely empty of any other vehicles. I knew if the temptation arose and I succumbed to putting my foot down heavily on the accelerator, the needle on the fuel gauge would drop quite dramatically.

I felt utterly frustrated, but I had to keep to the advised optimum speed of fifty-five mph; otherwise, I would not make it home to Llanelli or to be more precise, Burry Port. Bernard and I had agreed to stop at the Frankly Services on the M5 for a meal. Negotiating through the M6 /M5 interchange proved to be a complete dawdles, with hardly any vehicles around to hinder or encumber my journey. Shear bliss, an unheard of scenario, the infamous Spaghetti Junction section around Birmingham completely barren of traffic. Totally and utterly surreal.

We met at the Frankly Services, also deserted due to the lack of travellers, and ate our late lunch. Bernard's fuel situation on the van looked good and he remained confident he could make it back to the incinerator facility and near his home without any difficulty. My fuel situation also looked healthy; I estimated, provided I continued driving at a sedate speed, I should make it back to Burry Port with some fuel to spare. However, getting to the office and returning home the next day, a Friday, could be a completely different proposition. After leaving the services, I discovered the M5, much like the M6, to be completely deserted. As forecast, I made it home safely, finally arriving home that evening, although far later than I would have liked due to the low optimum speed I had to endure to complete the journey.

The next morning, with less than a quarter of a tank

of diesel I travelled the fifty miles to the office. Whilst driving past a local garage, I noticed a tanker off-loading some fuel, so stopped to replenish my empty fuel tank. The garage rationed me to three gallons of diesel, sufficient to get me to work in the morning and return home again that evening.

That was not my final experience with an acute shortage of fuel. A few weeks later, Steve Williams and I collected some hazardous waste from Smith Kline Beecham near Epsom. We had to take the waste to the large disposal facility which the major company held in Bedfordshire. Because we were carrying a few cylinders of Lithium Aluminium Hydride, a highly reactive combustible water reactive material, we were not permitted to transport the material through the Dartford Tunnel.

Instead of taking the much longer way clockwise around the M25, we decided to drive straight through London. While driving across the big smoke, one radio announcement made on one of the regional private radio stations alleged another embargo and blockade of oil refineries was about to start. Almost immediately, once again, motorists started to panic buy fuel, with cars queuing to fill up at garages. It became virtually impossible to travel through the capital as the traffic became snarled up and completely gridlocked around fuel outlets. The BBC desperately made regular announcements on all their national radio stations refuting the story.

Yet, despite all of the BBC's protestations, the queues on the garage forecourts became ever longer, as if the denials concerning the blockades at the refineries were complete fabrications and lies. Instead of stemming the tide of panic buying, the denials perversely exacerbated the situation. Consequently, it took us hours to negotiate through the city with immense queues of traffic agglomerating outside most garage outlets. When we did finally make it onto the M1 and started heading north, all we could see were the forecourts on the services and garages, inundated with motorists, desperate to completely fill their vehicles with fuel.

Our van also began running short on fuel because of our slow journey edging through the capital which rapidly depleted our diesel. If this continued, it began to appear we would be unable to reach South Wales. We managed to drop off the chemicals at the disposal site, quite late in the evening. We decided to cut across country and luckily for us came across a small garage which seemingly had a plentiful supply of fuel and appeared to be unaffected by the rumours of oil blockades on oil refineries and storage facilities and filled up with diesel. The garage manager had not even heard about the misleading allegations of any new embargo. The blockade story was completely unfounded and fortuitously completely fizzled out by the next day. I am ashamed to admit it, but the rumours emanated from an independent Welsh radio station based in Cardiff.

Those two incidents situation showed me how normal life could be completely disrupted and brought to a standstill simply due to a lack of fuel. It is a frightening prospect of a dystopian society which possibly lies ahead of humanity in the future, when the oil reserves do ultimately dry up. So be warned!!

CHAPTER 16

There are times in one's life when a personal desire and need arises to perform philanthropic, altruistic deeds. Such noble events have happened to me a few times in my life, instilling a feel-good factor about oneself, giving a warm glow of self-satisfaction. Knowing you have helped another fellow human being in some small way through this journey we call life. There is one specific incident I am still able to recall with complete clarity and lucidity; even though a fair number of years have passed since the episode occurred.

For this event, as the phraseology goes, we must *begin at the beginning.* The setting, a small Welsh village called Ferndale. Its predominantly terraced houses precariously perched on the sides of the Rhondda Fach Valley, encapsulated in the heart of South Wales. There, situated at the bottom of the valley, and within close proximity to the main street, was an old abandoned Victorian, red-bricked building.

One day, the local Fire Brigade received an emergency call to attend a blaze in the structure, and which occurred after some teenage children broke into the ancient, deserted building; deciding to keep themselves warm by starting a small fire utilizing some discarded, old wooden chairs as fuel. Unfortunately, the fire spread rapidly and quickly became out of control. The children swiftly fled the scene, but fortunately, a passer-by, using the oft heard media phrase, 'walking his dog,' observed smoke billowing from the building and immediately made a call to the emergency services. The Fire Brigade quickly arrived on the scene, broke into the locked building and speedily extinguished the small inferno.

After dealing with the blaze, the firemen became concerned about the contents secreted away inside the dilapidated, old building. For there, inside the structure, carelessly strewn about, they discovered an eclectic mix and concoction of toxic chemicals, together with deep, unfenced, sumps and tanks. The tanks each contained an assortment of sludges and stagnant water. The members of the emergency service experienced immense feelings of relief that none of the miscreant children came to any harm during their mischievous activities. They also realized the place posed a real danger, immediately informing the local Council and Environment Agency of their horrific discovery.

After investigation and a bit of digging by the authorities, they ascertained the building had once been occupied by a small company involved in all sorts of electroplating. Evidently, the owners of the company, after finding themselves in deep financial difficulty, did a moonlight flit, vacating the premises and abandoning all the toxic chemicals necessary for their electroplating operations. The owners also abandoned and deserted their workforce, obliging them to fend for themselves, leaving them unemployed, with unpaid wages and without any form of redundancy payment whatsoever. Unfortunately, this form of immoral, reprehensible and unscrupulous behaviour occurs frequently in business and the manufacturing sector, with companies taking generous grants, squandering the money, and then suddenly disappearing. Even huge banks perpetrated nefarious activities in the new millennium, gambling disastrously with investors' money, having to be bailed out by the taxpayer.

The Environment Agency, together with the Local Council, immediately decided the old building should be cleared of all the chemicals as quickly as possible. Together, they began approaching various hazardous waste disposal companies. Of course, amongst those specialised companies approached was Chemkiln.

Rupert and I travelled to the site for the initial inspection and were also horrified and appalled at what we discovered inside the old building. The costings proved difficult due to the number of imponderables and unknowns inside the premises, having to be worked out on a worse-case scenario. Yet, despite the quoted price for the project being quite high, our company won the contract.

A brief time later, we made a start at decontaminating the building. Sealing it off and putting up the regulatory safety and warning signs. The company allocated me Project Manager for the task, with three other team members working under me. Our first task was to empty all the sumps and tanks, after first taking samples to determine whether the sumps contained neutralised chromium hydroxide sludge, as I surmised. I knew my time working in the electroplating industry would eventually pay off and prove invaluable. My assumptions concerning the contents of the tanks proved to be correct. I also ensured the tanks and sumps contained no cyanide, one of the chemicals often used for electroplating.

Additionally, I determined if any other toxic chemicals were present in the sludge for the necessary legal documentation required for shipment and disposal. If any items on the documentation proved to be incorrect, then I, along with Chemkiln, could be prosecuted, for the legal documentation relating to the waste, bore my signature on it. I calculated the volumes of the sludge contained in the sumps, utilising basic arithmetic, after taking the dimensions. Once I had ascertained the volume and contents, I ordered road tankers to take the sludges and stagnant water to various authorised water treatment sites. Treatment works being the preferred option and better suited cost-wise for disposal, rather than by the more expensive incineration.

During the initial days of the project, I often observed a man, who I estimated to be in his mid-thirties, watching from outside the perimeter fence. The next time he stood by the fence, I decided to approach him and find out what he wanted and why he appeared to be so interested in all the activity. So, at the first opportunity, I walked over to him and began chatting. Following a discussion, I discovered his name was Richard. It appeared he had worked for the last owners as a supervisor. Because of this association, he was totally familiar with the site.

The moonlight flit by his employer eighteen months earlier came as a complete shock. Unsurprisingly, the whole situation had made him slightly bitter and resentful. Since that time, he had only managed to obtain, limited, casual work. At that instant in time, he remained unemployed, having to support a wife and two young children. My heart went out to him; he had a young family to support, with Christmas rapidly approaching. Unfortunately, job opportunities in the South Wales valleys, at that time, tended to be extremely limited. He asked if he could come inside the building and look around at what we were doing.

At first, I resisted, but he stressed he would be very careful, and, after all, he had once worked there and knew the building layout and the inherent dangers. I reluctantly agreed, only after first supplying him with all the necessary safety equipment. As he slowly walked around the old building, I accompanied him, introducing him to the General Services team. The team gave him cursory acknowledgement while they went efficiently about their work.

Whenever we came upon piles of unmarked chemicals, Richard would enlighten me as to their chemical composition; sodium hydroxide, potassium cyanide, chromium oxide, potassium hydroxide, sulphuric acid etc. The worst he mentioned, and one I had never personally come across during my time in industry, cadmium cyanide. The cadmium could not be incinerated, being a red listed element and an extremely toxic metal, which would vent to the atmosphere. The other chemicals tended to be run of the mill chemicals used in electroplating.

After Richard left, I took some samples of the materials for analysis by the plant laboratory at the incineration facility. When the results eventually came back, they corroborated exactly with Richard's information, which proved to be spot on. From then on, I had complete confidence in his knowledge of the site. Whenever we came upon additional unmarked chemicals, if, upon observing Richard walking his dog nearby, I would generally approach him to pick his brains. Richard's help proved to be invaluable. His knowledge of the unmarked chemicals on the premises helped to speed up and expedite the project considerably. If only there was some way in which I could repay him?

The project lasted a few weeks. Sometimes we had to leave it, for example awaiting collections by outside haulage waste disposal contractors, one of which was Hyperwaste, my old employer. This gave me a chance to chat with the Hyperwaste drivers and catch up with events concerning the company and its personnel. As mentioned, some of the waste went for treatment, some for landfill, and some for incineration. The disposal route was completely dependent upon the chemical constituents.

As the project gradually approached its finale, I drove there straight from home, to meet the team to carry out some tidying up at the site. While driving there, large flakes of snow began slowly falling, surreptitiously agglomerating on the road. I drove gingerly along the Heads of The Valleys road to the small town of Aberdare on the way to Ferndale, negotiating through the snow enriched town, then up the steep, white mountain road leading to Maerdy, a village en route to Ferndale, also in the Rhondda Fach valley. My Rover 75 slid and slithered up the narrow mountain road as it desperately struggled to obtain traction on the icy surface. What should I do? The falling snow appeared unrelenting, Ferndale is not the sort of place a person would wish to be marooned during a snowstorm or blizzard.

As far as I could recall, the small hamlet possessed no hotel in which to shelter for one or possibly two nights. Heroically, or foolishly, dependent upon your stance, I carried on regardless, heading on up the mountain road. From near the top, the view was breathtaking, with the panoramic scene of the countryside below covered in a blanket of virgin snow. The view obstructed somewhat by the falling flakes and the misty, overcast, gloomy conditions.

I persevered, onwards and upwards Macduff. After reaching the top, the car slowly descended into the smaller of the two Rhondda valleys, first through the small village of Maerdy. I encountered very little traffic on the mountain road, ironically, making driving relatively easy. After what seemed like an interminable amount of time, I finally entered Ferndale, its houses enshrouded in powdery snow. I had yet another decision to make, should I drive to the site.

With the building position at the bottom of the valley, it would entail driving down an exceptionally steep incline, if I drove down the incline, there was the distinct possibility I would not be able to drive back up it. To hell, with it! In for a penny, in for a pound. I drove, or rather glided down the incline, finally sliding my Rover 75 towards the site. The weather reports on the radio indicated the snow storm would quickly blow over. I held my faith in that. Mercifully, it was not Michael Fish reporting the weather like 1987, so I felt reasonably confident in the Meteorological Office's prognostication. I phoned Bernard informing him about the conditions and told him he did not have to come to the site if he did not want. He answered he would try to get to the site and leave shortly. I began pottering about sorting the chemicals ready for collection and disposal making a thorough, detailed inventory of the remaining chemicals.

After about two hours, I heard Bernard drive up in the company van. The two of us worked for a few hours, by which time the snow stopped falling and the sun miraculously made an appearance. During the afternoon the weather gently improved and the day became warmer. Consequently, the snow gradually began to thaw. By late afternoon the roads had cleared. The only snow left to indicate there had ever been a snow storm, lay in isolated pockets, being shaded from the sun allowed the isolated pockets of snow to survive longer.

In the final days of the project, quite a bit of scrap metal remained, the metal lying about; all of which had to be removed from the site. We had been told as a sort of bonus, any clean, uncontaminated metal could be sold for scrap and the proceeds split amongst the team. I contacted a local scrap dealer and struck a deal, whereby he took all the uncontaminated metal on the site. By the end we received almost two thousand pounds from the deal and I decided we should give Richard some of that money. His help had been invaluable. His assistance had saved the company a lot of time and money and I felt it only fair and morally justifiable to recompense him for his assistance. On the last day everything had been removed and the last bit of uncontaminated scrap collected and the building was completely empty and safe. With every one of the General Services team who had been involved in the project on the premises, I called them all together, informing them of my decision to give Richard a cut from the scrap fund. Only one person dissented; the curmudgeonly Albert.

'Why should he get a cut?' he protested angrily.

'Because his help saved us a lot of time and made the job easier. I think he deserves it.' I argued.

In all fairness, the rest of the team agreed with me, also considering it the right and proper thing to do.

After that, we took a vote agreed to my suggestion by a majority decision. True Democracy at work, while Albert sulked and pouted at being overruled and being deprived of some extra cash.

Richard had given me his telephone number before we vacated the old building for the last time. I phoned him and asked if he would come down to the site on the pretext of needing his advice.

Shortly afterwards, Richard appeared in the building. With all the work finished, the team stood idly around only waiting to head back to the incineration facility. Upon observing Richard, I approached him. Earlier that day, I had meticulously counted out five hundred pounds from the wad of dirty notes which had been given to me by the scrap dealer, adding another hundred from my own cut.

'Hi Richard, thanks for coming down at such short notice.' I said.

'The boys and I have made some money from selling some uncontaminated scrap metal from the site. We have had a discussion amongst ourselves and we all agree, as a *"thank you"* for all your help over the past few weeks, we think you deserve a share.' I then pulled out the wad of segregated grubby notes from my pocket and handed it to him, then adding, 'Merry Christmas.'

The team all chirped in as well, apart from Albert, that is, who remained sullen and quiet during my short speech.

'Merry Christmas, Richard!' they shouted in unison. The look on Richard's face became a picture to behold, emanating a mixture of surprise, bemusement, happiness and emotion.

'Thank you. I don't know what to say,' he croaked, the emotion evident in his trembling voice.

'You don't have to say anything,' I responded.

'The look on your face says it all. We all hope you and your family have a happy Christmas.'

At that instant, I thought I detected him welling up, with the slight hint of tears exuding from both his eyes. 'We must go now,' I said and shook his hand for the last time, adding, 'and once again, thank you for all your help.'

Richard returned home to give his wife the good news. After locking the premises up for the last time, all the team departed, with Albert still sulking.

Shortly afterwards, we all headed back in our assortment of company vehicles, back to the incineration facility. I drove alone in my Rover 75. The day was cold but sunny, with not a cloud in the sky, which helped elevate my spirits even more.

But first, I had to return the keys, which locked the building back to the local council offices. While driving along, a warm glow came over me, as I heard the unmistakable, gravelly voice of Louise Armstrong singing. 'It's a Wonderful World,' emanating from the radio. With the uplifting words reverberated through the car speakers. I allowed myself a self-satisfied smile, trusting Richard and his family would have a far better and more enjoyable Christmas than they had originally anticipated.

CHAPTER 17

*F*ollowing my almost year-long, sexually stimulating affair with the libidinous Rebecca, I did not even contemplate embarking upon another relationship; through the internet or by any other means, but simply intended concentrating upon my career with Chemkiln, and metaphorically speaking, try rebuilding a few of the devastated relationship and emotional bridges with Stella. This, I discovered, turned out to be an extremely difficult, frustrating uphill task. After only a few months, mainly because of immense communication difficulties between Stella and myself, I succumbed to the temptation of another love affair. Although, in mitigation, the new affair took me totally by surprise, emerging completely out of the blue, catching me entirely unawares, as many affairs and relationships have an unnerving tendency to do.

While at home one evening, searching the internet for some obscure technical information relating to work, I swiftly became bored, deciding to venture into one of the many AOL chat rooms which seemed to proliferate the internet at that time. Suddenly, in one of the rooms I came across a solitary internet user. To be more precise, the person came across me. After '*chatting*' for a while, I discovered the lone person to be a recently estranged married woman in her early forties, and in the process of going through an evidently acrimonious divorce.

I also quickly ascertained it to be her very first evening on the internet, and her initial experience of a chat room, a new inductee to its wonders, having only installed her newly acquired computer and its associated gubbins that very afternoon. She described herself as being a complete novice, and seemingly, not very computer or internet literate. Discovering her name to be Bridget, I also gleaned from her that she had two young, pre-teenage sons, Billy, ten, the elder of the two by eighteen months, and Jack. Bridget and her two sons lived about fifty miles away near in a small, picturesque fishing village, located just up the coast from Llanelli in South West Wales. After an hour or so, we both agreed to end our discussion and log off. I believed, erroneously, as it happened, our initial encounter would probably be our last. '*Ships that pass in the night,*' and all that.

The following week I was on my travels with work and for one reason or other, unable to access the internet. During the last decade of the twentieth century, Wi-Fi and broadband were both technological concepts of the future, with the burgeoning internet still in its infancy. Due to the technological inadequacies at that time, it would eventually be a couple of weeks before I had the time and accessibility to the internet at home, and able to venture into the electronic highway, to visit one of the many chat rooms. The first evening I could access the internet, I entered one of the chat rooms. My newly acquired friend soon made herself known to me, and we soon became immersed in another prolonged internet chat.

This scenario continued for several weeks, during which time, at Bridget's request, I sent photographs of myself. Although in all truthfulness, they had been taken a few years earlier, when I was far younger, with my features exhibiting quite a few less wrinkles and age lines. Come on, who doesn't do that? We've all done it? Yet Bridget seemed slightly reticent in reciprocating by e-mailing pictures of herself. Not that I was all that bothered, for the way she described herself as being adorned with long, unkempt shaggy, vibrant red hair with various coloured streaks running through it, accompanied by a stud pierced nose, in addition to the numerous tattoos which apparently liberally adorned her body.

All these bodily decorations portrayed and illustrated by Bridget helped dispel any intention on my part of us ever meeting up or ever becoming romantically involved. From the portrayal which she gave of herself, Bridget appeared to be adorned like a punk. I personally consider punk to be a disgusting, aggressive period of British pop culture and must confess to being absolutely delighted when that period in pop history eventually fizzled out sometime during the eighties.

I didn't realize it at that time, but Bridget was stalking me for ulterior motives, motives which would eventually become apparent much later. But at that time, the intense interest she appeared to be taking in me massaged my fragile ego. During our chats, I informed her about my personal life and the increasingly strained relationship evolving between Stella and myself. One evening, after unexpectedly divulging her telephone number, Bridget asked if I would phone her. Reluctantly, I acquiesced to her request. I discovered myself pleasantly surprised.

Despite all my pre-conceived, somewhat snobbish ideas about her; Bridget sounded extremely articulate and tremendously well-educated; possessing a gorgeous, husky sexy voice. As the reader is aware by now, I am a sucker for the sound of a sexy female voice, finding it a distinct turn on and quite an aphrodisiac. Consequently, despite my good intentions about not embarking upon another affair; in the wake of that initial telephone conversation, I found myself well and truly hooked, unawares of slowly being reeled in. From that time onwards, our verbal discussions over the phone became a fairly regular event and inevitably more risqué.

During the next week or so, I had been booked to attend a two-day management course, which necessitated an overnight stay in Birmingham. Bridget suggested we meet up, with it being the week of her forty third birthday, and she wished to celebrate it with me. I had still not seen any pictures of my internet friend and became thoroughly intrigued as to what she actually looked like. So, contradicting and completely ignoring all my self-imposed reservations about embarking on another affair, I acquiesced to meet her for a meal one evening.

Throughout the tenure of the management course, I stayed in a medium sized hotel located in the Hagley Road on the outskirts of Birmingham. Bridget suggested we meet at a restaurant which she knew of, located on the outskirts of Ross-on-Wye, a small town located on the England, Wales border, where Dennis Potter, the famous playwright, once lived.

The night arrived, and I quickly left the management course, heading back to my hotel. After a quick shower and change of clothing, I phoned Bridget, informing her I could meet her, as I had no work or projects to do that evening relating to the course. We agreed the place and time to meet, allowing Bridget sufficient time to drive from her home in West Wales.

Being the slightly closer to the agreed rendezvous location, I had plenty of time to spare, eventually leaving my hotel some forty-five minutes later, arriving in the car park of the restaurant about seven pm, some thirty minutes earlier than agreed. While I awaited the arrival of my unknown, yet to be visualized, slightly enigmatic date, pleasant congenial memories of my first encounter with Rebecca, and our meeting in the car park of the Ibis hotel in Swindon, came flooding back. Circumstances seemed very similar to that meeting all those moons earlier. Being late October, it was also a dark evening, and the weather slightly overcast. Bridget ended up being very late for our assignation, turning up some forty-five minutes after the agreed time. Well past eight o'clock.

If Bridget had turned up a quarter of an hour later than she finally did, I would have probably been returning to my hotel in the Hagley Road. She did not have a mobile phone, so it was not possible to contact her to determine her whereabouts. Finally, just in time, her white Citroën Saxo drove into the car park. After parking it next to mine; she looked directly towards me, at the same time exuding a wonderful smile. I must confess, at first sight she did not set my pulse racing like Rebecca had done almost a year earlier. Bridget was not unattractive, but certainly better looking than I had originally envisaged in my mind. Yet, despite her reasonable charming looks, she just did not arouse my animal passions.

Bridget turned out to be of slim build, about five feet four inches tall, with a nice slender figure and shaggy red hair, with coloured streaks running through it. Her red hair had been obviously generated through the aid of chemical supplements. Also helping maintain the description which she gave of herself, Bridget exhibited the small silver stud at the right side of her nose, and she was attired in a long, but wearing a hippy-like, flowered dress, not leathers or jeans, which I had anticipated. Her tattoos were not visible, having been etched onto parts of her body masked by her clothing. From her outward appearance, together with her long flowing dress, Bridget could very easily be ascribed with the tag of *tree hugger*.

After emerging from our respective cars, I gave Bridget a formal peck on the cheek. We then made our way into the restaurant. It was a Monday evening, so I had not made a reservation, believing there would be no difficulty in obtaining a table for two people in the early part of the week. My assumption proved to be correct in that respect, with the restaurant virtually empty, displaying plenty of vacant seats and tables. We sat at a designated table and began talking. Initially

Bridget kept apologising profusely for her lateness.

Being fully aware of Bridget being a vegetarian, information, a fact gleaned through our internet and telephone discussions, I ordered a vegetarian meal for myself. The decision turned out to be a huge mistake on my part. The meal obtained the unfortunate distinction of probably being one of the worst meals I have ever experienced in my life. The lettuce was far too cold, and it did indeed feel like consuming an iceberg. Iceberg being an apt name for the vegetable. I did not enjoy the meal one iota. While I struggled through the dinner, Bridget told me about herself and her two sons, informing me of her passion for animal rights. She had also once lived in Florida and New Zealand, learning to drive and obtain her licence there, when just sixteen, and where she had also obtained a pilot's licence. She seemed to be a well-travelled, interesting person. Unfortunately, unlike the assignation with Rebecca, for some inexplicable reason, Bridget simply did not push my emotional or lustful buttons.

At the end of the evening, we had a bit of a snogging and petting session, but that was all. Because of her family commitments with her two young boys, Bridget had to get back at a reasonable time, allowing the babysitter to also leave at a respectable hour. Prior to heading back to my hotel on the Hagley Road, I needed to put some diesel in my car. Bridget accompanied me to the garage in her car. Before I left, she gave me a long, passionate kiss. It was like being in a movie. Bridget lifted her right calf, at the same time as she gave me the slow, lingering, passionate kiss, then whispering seductively in my left ear.

'Thank you for a fantastic evening. I've had a wonderful time.' We then both got into our cars and headed our separate ways.

I thought that was the end of it. However, the next day, I had a frantic telephone call, it was Bridget asking if, after the course, during my journey home, whether I could stop at one of the garages near Monmouth, where she had left her purse, then, would it be possible to take it to her home. I reluctantly agreed as the expedition to her home would mean adding at least an extra one hundred miles to my trip home. After taking the details of the garage and her home address I set my mind on the getting home quite late that night. It didn't matter as Bridget would be in work.

Thankfully, when I did eventually reach Bridget's home that evening, there was no answer, so she was evidently out. I had no intention of staying for a cup of coffee and quite relieved no one was at home. Pushing the purse and its contents through the letterbox, I headed home.

Bridget contacted me the next day, via e-mail. Thanking me for my '*Good Samaritan*' act and asked if we could meet up the following Saturday evening while Stella was at work, she promised to buy me drink and thank me for dropping off her purse. '*What the hell?*' I thought, agreeing to her suggestion.

We met at Resolven, a small village about halfway between our two houses and went to the Red Lion, one of the pubs located there, for a drink. Bridget looked gorgeous. She had evidently made every effort to impress and looked absolutely stunning. It did the trick, as my hormonal juices, started to flow. I am only flesh and blood after all.

This time, our assignation involved no meal, and we both had low alcohol or soft drinks, because of driving our own cars, both of us adhering to the drink drive laws. We located ourselves next to each other in an isolated alcove with Bridget sitting quite close to me continually rubbing her left leg against my right, while at the same time gently caressing my hand. During the conversation, she informed me that she had recently been having an affair with a married man, but she had ended it that week, considering there to be no future in their liaison and now wished to have relationship with me. Although I could not fathom the logic, for I too was in a sort of relationship, so why should the situation be any different in that respect? While we talked intimately, she snuggled ever closer.

'Do you know of a secluded place nearby where we could be alone without fear of being disturbed?' She whispered sexily into my ear. Seduction and lust appeared to be most definitely in the air, Bridget was making her shameless intentions unequivocal.

Being well acquainted with the area, I informed her of a secluded lay-by I knew of, which was well away from the main road and might suit our licentious purposes.

'Well, what are we waiting for? I'll follow you?' she responded

With that, we quickly consumed the remains of our drinks and drove in our separate vehicles to the designated lay-by.

Fortuitously, the place was deserted when we arrived there. I parked my car and Bridget parked her small Citroën Saxo behind. After locking the doors to her vehicle, she walked to the side of my car before virtually ordering me into the more spacious back seat. The instant we were both in the rear of the car, Bridget dived onto me and then proceeded to French kiss me passionately kissing open mouthed pushing her lips tightly against mine, 'I've been aching to do this all week.' she purred huskily into my ear. Before I had time to regain my breath and say anything, she began undoing the buttons on my shirt then unclipping the buckle of the belt and releasing the button securing my jeans and before seductively pulling down the zip at the front.

'I am going to give you an evening to remember,' she promised, continuing to purr. With that she moved her head down my body, firstly kissing my chest then my stomach before reaching my underclothes. After pulling down my underpants it was evident she had indubitably succeeded in arousing my passions.

'Well somebody is most definitely pleased to see me,' she joked. Then she proceeded with more experienced sexual manipulations. I was in seventh heaven; this was absolute female seduction at its very best.

After a while, Bridget finished her sexual ministrations pulled up her long, dress and straddled me. No underwear was involved to impede proceedings and we consummated our relationship. She was virtually raping me. Perhaps rape is the incorrect verb, for this was willing male surrender, capitulation and participation. She moved up and down rapidly before we both simultaneously reached our climax and satiated pleasure.

She lay quietly on top of me for some moments, her head resting on my chest. Finally, breaking the silence, she spoke.

'I sure needed that,' she uttered finally.

'You needed it?' I rhetorically replied, imparting the information I too needed the sexual release.

We talked for a while, before I had to get home just before Stella came home from work. After kissing we parted company and from that night onwards I was caught like a fly in a spider's web, emotionally and sexually trapped, Bridget had completely and utterly beguiled me. I had been well and truly reeled in like a game fish.

Some weeks passed by, after that encounter. The major holding company had acquired a chemical cleaning company in its ever-increasing corporate empire. The newly acquired company was very short on highly technical people for the chemical cleaning and required the services of a supervisor/chemist for the technical analysis. Somehow my name began to appear in the memos and e-mails flitting around the company.

Dave, our new head of department, approached me one day and asked if I would be interested in helping Atlas Clensol with the technical side of the operation after they had managed to acquire the tender for the project for passivating stainless-steel pipework at Astra Zeneca in Loughborough. I knew the facility as we went there regularly disposing of laboratory chemicals. I thought a change of duties would be good and a subtle change. As the saying goes, 'a change is as good as a rest,' applying and it had been quite a while since I had been involved in chemical cleaning operations with Hyperwaste. It would be nice to keep my hand in. The project had been forecast to last for two weeks and would mean being away from home for a couple of weekends.

During one of our chats I told Bridget about the project. She appeared to become very excited and told me she had some friends who lived nearby in Leicester also informing me her two sons would love to stay with them and would it be possible for her to stay with me during one of the weekends.

I considered the request carefully then thought why not? We could spend the weekend together and the guys working for Atlas Clensol had no idea what Stella looked like; I could easily pass Bridget off as her. Besides, they were all staying at another hotel and they would, probably, never meet her.

We both agreed Bridget would travel to Leicester on Friday. We would then meet up on the Saturday at the Great Western Hotel Loughborough. Friday night came, and I had arranged to meet the team of guys from Atlas Clensol for a drink and curry at a nearby restaurant which was located about half way between their digs and my hotel. While I was walking out of my hotel, a small white Citroen pulled up in the car park. I could see the car contained Bridget and her two young sons, Billy and Jack. This was the first time I had laid eyes upon them and was completely shocked to see them. Bridget parked her car, got out and rushed towards me.

'I couldn't wait until tomorrow, I had to see you today,' she said before planting a passionate kiss on my lips.

I was absolutely dumbfounded this was not what we had agreed; still, it was nice to see her again. She then introduced me to the two, her two sons. I told her I was about to meet the team for a curry and would have to phone them to say I was ill or some other excuse, which I did at once.

I asked if they had all eaten, which they had not, so I took them to a nearby McDonalds for a happy meal and which Bridget suggested. After which we went to my hotel room for short while where I quickly became acquainted with the two boys and we talked for some time. It was impossible for them all to stay in my hotel room, it was far too small, so they had to head to Bridget's friends before it got too late.

The three of them left and I arranged to meet Bridget at the hotel the next evening at about five pm after work. We had a lot to do that Saturday at the Astra site, but it would give Bridget the opportunity to take her two sons around Leicester.

Plans can invariably go awry, and the work scheduled for that Saturday was not fully completed until seven in the evening, and I did not get back to the hotel until seven thirty. Bridget was waiting for me in the lounge and during her sojourn had consumed a fair few drinks and consequently, slightly tipsy.

After a quick shower and change of clothing, we had a meal in the hotel restaurant and talked for quite a while. Eventually as Zebedee from the Magic Roundabout would say, 'Time for bed.' And we made our way to the room. Both of us could not wait to undress and jump naked into bed and cuddle up to each other. That night passed as a blur while we indulged in passionate, exhausting lovemaking.

That Sunday morning, we did not have to go to the site, so Bridget and I arose at shocking time and made our way to the restaurant for a late breakfast which we only just managed to make before the breakfast bar closed. The waitress had served us the previous evening and smiled at us both, although it appeared to be more of a smirk and I thought I heard her jokingly remark to one of her colleagues.

'She stayed with him all night. Randy old sods. You would have thought they were past it at their age.'

We went for a walk around one of the local parks before Bridget had to go and retrieve her two young sons and head back to West Wales.

Our relationship continued to flourish. I often went down to her house, adding the time on to my work schedule, with the black hole of time as I had previously done with Rebecca. Bridget often phoned me while I was at work, which tended to be a bit awkward with my colleagues often being nearby.

Towards Christmas 1999, Chemkiln obtained the contract at Queen Charlotte's hospital for the removal of a large amount of toxic chemicals. As an aside, I never knew there was a Queen Charlotte, but soon discovered she was the wife of King George III. All the female consorts of kings are given the title of queen. The project once again was a long drawn out affair and half way through the contract, Bridget phoned and said she would like to travel to London visit a cousin of hers and stay the night with me. As with Rebecca and her trip to Jersey, I was ambivalent about the situation. Sure, I would like to spend the night with Bridget, but once again, I did not want my colleagues becoming aware of our relationship.

There were three of us working on the project, my old friend Bernard Evans, Steve Williams and myself. How the hell was I going to keep them from meeting Bridget? We were staying at a small hotel located between Ealing and Acton and virtually on the border of the two boroughs. The night before Bridget was due to arrive by train; I had a meal with my two colleagues at an Indian restaurant in Ealing. During the evening, I casually informed them I would not be joining them the following evening but intended going to the cinema instead. Planting the seeds, I surreptitiously suggesting they should go to Acton and have a meal in one of the many ethnic restaurants, which flourished in that area. Both my colleagues gave every indication they would follow my suggestion and do just that.

The evening of Bridget's arrival soon came around and after completing our day's work, our work at Queen Charlotte's hospital, I immediately retired to my hotel, phoning Bridget who had by this time acquired a 'pay as you go' mobile phone. She informed me the western line train had just stopped at Swindon station and anticipated being at Ealing Tube Station around seven thirty.

I felt extremely nervous, hoping I could keep Bridget from meeting Bernard and Steve, desperately wishing to keep my sexual tryst a secret. I had no desire for my associates to have any knowledge of my secret liaisons knowing full well how they would rib me mercilessly. To take my mind of how things, I had a long soak in the bath to expunge the sweat and grime of the day and make myself smell acceptable for Bridget.

The tube station was not too far from the small hotel and I had told Bridget to phone me when she arrived at the tube station. At seven forty-five, my mobile phone rang, it was Bridget; informing me she was at Ealing Station and would wait for me to meet her.

I was ready and quietly vacated my room, hoping

not to bump into Bernard or Steve. Unfortunately, Steve's room was next to mine hence my reason for being as quiet as possible. Luckily, I did not see either of them as I made my way to meet Bridget. I soon arrived at the temple of the London commuter and observed my new lover standing at entrance. When she saw me she immediately exuded a broad smile and hurtled towards me, grabbing and then kissing me passionately.

'I thought I would never get here!' she exclaimed.
'Neither did I. Are you hungry? I thought we would go into Ealing for a meal,' I suggested.
'Sounds fantastic, shall we go?'

I grabbed Bridget around the waist and we walked the relatively short distance to Ealing High Street. I was comfortable, believing, at that instance, my associates were wending their merry way to Acton, Completely in the opposite direction. I asked Bridget what type of meal she wanted her answer indicated a vegetarian salad would be ideal. I knew of a pub which would be ideal, providing vegetarian meals and so we headed towards it. Imagine my complete horror upon entering the pub only to observe Bernard and Steve sitting at one of the tables. Steve had his back to me, but unfortunately, Bernard was looking directly at me. There was no way we could exit now. We had been well and truly found out or in the modern-day idiom, 'busted.'

'Oh my God,' I muttered to Bridget through pursed lips, 'there are Bernard and Steve, my two work colleagues. We are going to have to sit with them. I can't just completely ignore them.'

To my astonishment, Bridget appeared completely unperturbed and looked as if she was thoroughly enjoying the whole, farcical, embarrassing situation. What she didn't realise was, Bernard knew about Rebecca, my other secret liaison, but he had never actually met Rebecca during her short visit to Jersey. I made sure the two of them never met. He was bound to believe Bridget to be Rebecca. Also, Bridget had no inkling about Rebecca and that pleasurable episode in my life, and which I had kept from her. As far as I was concerned this unexpected turn of events was about to become very awkward and extremely uncomfortable. Well, certainly for me at any rate. As we approached the two of them, I spoke before anyone else had a chance to say a word. Looking Bernard directly and intently in the eye I began the introductions.

'Bernard, I would like you to meet Bridget. Bridget this is Bernard.' Every time I said Bridget, I deliberately emphasised her name. Oh, what a tangled *web we weave*.

I then looked at Steve. 'Steve this is Bridget, Bridget this is Steve.'

Through the corner of my eye, I observed Bernard, once again, gently shaking his head in disbelief and incredulity. I had said her name so often, there was no way he could have failed to get the hint. Whether Bernard had informed Steve, about Rebecca, I had absolutely no idea. Although knowing Bernard, it was a fair bet he had told everyone within the department about my secret assignation in Jersey. So, there was always the distinct possibility Steve could also let the cat out of the bag.

We all had a meal together and I was on tender hooks as to what would be said by any one of them. Fortunately, as far as I can recall, nothing untoward was said during the meal. I could not wait to finish the meal and leave the pub. Wishing to be alone with Bridget was one part of the equation. But the main reason was to get away from Bernard and Steve as quickly as possible and put an end to this shear purgatory.

Bridget and I left the pub as soon as we had finished

our meals and headed back to the hotel. We got into the room and Bridget smiled at me.

'That was unexpected and fun!' she exclaimed excitedly.

I did not say a word, fun was not how I would describe what had transpired and just kissed her. Soon afterwards we were in bed and making love.

In the early hours of the morning, Bridget resumed her carnal desires and continued the assault on my body once again. Unfortunately, as mentioned Steve's room was next to mine and I also did not realize the adjoining walls were not that thick, subsequently exhibiting very little soundproofing. Accordingly, Steve could hear most of the noises emanating from my room during the night, allowing him very little sleep.

The next morning, Bridget insisted on going down for breakfast. It was then Steve made it abundantly clear he had not experienced much rest due to the excessive carnal noises radiating from the room next door. Unfortunately, during breakfast, also Bridget made it evident I was visiting her home on a regular basis. Which, hypocritically, Steve considered immoral, considering all the help I gave him with the married Cheryl.

We finished our breakfast with me still on tender hooks. Bridget left shortly afterwards to visit her cousin in Lambeth.

Our relationship continued to blossom and soon affected my life in ways I could not even comprehend at the time.

Unfortunately, my liaisons became known within Chemkiln. During Albert's leaving retirement party, Paul Morgan, one of the General Services supervisors who had left the department to work for the holding company in another part of the country, turned up, along with his long-time girlfriend Shelly. Shelly came up to me, grabbed me, before giving me a friendly, sisterly kiss on the cheek as a greeting. Almost immediately, Paul pulled her away.

'Get away from him!' he instructed her, looking at me as if I had suddenly displayed symptoms of leprosy. He then remarked to me,

'I've heard all about you. You've got quite a reputation,' which made me smile. Bernard had obviously been talking and exaggerating the situation somewhat.

Now, I never considered myself to be a Lothario when it comes to women. Far from it, in fact, turning females off, rather than on. It only goes to show how things can become highly distorted.

CHAPTER 18

Overall, and for the most part, many of the projects I became involved with working on behalf of Chemkiln, tended to be mundane, boring and relatively straightforward. This generally tends to be the case with repetitive, unvarying tasks, performed with whatever profession, career, job or vocation one happens to be involved with. However, having made that all-encompassing, fairly generalised statement, my employer did occasionally manage to obtain projects which fell into the category of being highly dangerous, fraught with many perils, completely out of the ordinary, exhibiting an extremely high-risk factor and verging on the terrifying.

We categorised and pigeon-holed these as the, '*Oh shit projects!*' Although, I must admit, from a completely personal aspect, I always experienced a certain buzz and exhilaration when faced with such enterprises; often undergoing an extreme adrenaline rush. One such project, which would ultimately prove to be the last major project I would supervise for Chemkiln, involved working for the Ministry of Defence (MOD). To be more specific, working alongside the scientists, engineers and explosive experts normally based at the highly secretive, clandestine research establishment located in the famous Porton Down. The specialised personnel possessed the requisite technical expertise to look after and maintain Britain's ballistic missile arsenal.

Ballistic or Cruise missiles comprise of two major chemical components. The first is the fuel, such as hydrazine, kerosene, ethanol, or some other highly flammable, combustible material. The other major component is the oxidizer. When these two chemical components are mixed together in the correct proportions; the resultant mix reacts violently, generating thermal energy. This thermal energy converts into kinetic energy. It is this kinetic energy which propels the missile or rocket, helping it to travel through the atmosphere or, as in the case of the Apollo space programme, into outer space. One of the liquid oxidizers often used in the British ICBM and cruise missiles is a highly dangerous material, given the abbreviation IRFNA (Inhibited Red Fuming Nitric Acid): a highly dangerous concoction of nitric acid, dinitrogen tetra-oxide, water and hydrofluoric acid.

For almost half a century, during the early years of the Cold War, the MOD amassed large stockpiles of this highly toxic, corrosive, oxidizing material, storing the liquid in aluminium containers. The part, which, to this day I still find incomprehensible, contradictory and utterly incongruous, is the fact the 800 litre aluminium containers had actually been manufactured and supplied by Russia. God forbid, had hostilities catastrophically erupted between the West and the Eastern Communist bloc, then paradoxically, the IRFNA stored in those very same Russian manufactured aluminium containers, would eventually help propel the cruise missiles against the Eastern Bloc, including Russia; indirectly assisting in decimating those countries and their populations.

During those highly charged, turbulent, unstable years of the Cold War, as part of its strategy, the MOD dispersed enormous quantities of IRFNA throughout the United Kingdom, preventing all the stocks of the oxidizer being destroyed in one fell swoop. Referring to the old well-worn cliché, '*not putting all of your eggs in one basket.*' If all the containers had been destroyed by a pre-emptive attack perpetrated by the Eastern Bloc countries, their obliteration would have negated and thwarted any possible large-scale retaliation by the British government of the day.

By the closing decade of the twentieth century, a completely different historical, political scenario prevailed in the conflict between the Capitalist West and the Communist East, the Cold War having thawed considerably. The Labour government and the MOD considered it a more fiscally prudent option to move and store the highly dangerous IRFNA in a few large storage facilities. After all, the likelihood of a pre-emptive strike by the Russians had, by this stage in the timeline of mankind, diminished markedly, from that of being a distinct possibility, reduced to a highly unlikely prospect. Hence, plans began to formulate for relocating most of the 800 litre containers dispersed throughout the United Kingdom. The intention, to store them in a secret ordnance and munitions depot, located in the south west region of Scotland. The location, once the biggest munitions factory built during the Great War, constructed on the orders of a great Welshman, Lloyd George.

Subsequently, one day in the first year of the new millennium, the MOD shipped over fifty of the aluminium containers, together with their lethal contents, to the undisclosed, aforementioned location in Scotland, arriving there in the early hours of the morning. Upon their arrival, the selected, detailed MOD personnel, immediately began off-loading the aluminium containers into one of the empty large buildings and which resembled a huge aircraft hangar. The personnel located the containers inside the building with military fastidiousness and precision, placing them side by side in neatly aligned rows. They performed the whole exercise covertly, with efficient, military exactitude, and in complete secrecy.

In fact, the personnel performed the operation so surreptitiously and proficiently, the private security firm assigned with the task of looking after the protection of the facility, had not been informed about the influx of the mysterious containers, and their lethal contents. Remaining completely ignorant and unaware of the containers, they only knew about the few MOD vehicles driving onto site, but not the actual shipment being transported.

During the whole operation, an incident occurred which went unnoticed. The episode transpired during the loading and handling of the old aluminium vessels, when one of the metal containers slightly ruptured. The long, tortuous, jarring road journey, exacerbating the situation. The minute split in the vessel, escalated into a small leak. At the other end, throughout the off-loading procedure, the small leak somehow went unobserved, with the military personnel placing the container inside the large building, oblivious to the impending doom.

For days afterwards, the leak worsened, with the highly dangerous, volatile contents surreptitiously oozing out of the container. The liquid vaporised, and the toxic brown, corrosive fumes slowly began completely filling the immense void inside, permeating and utterly dominating the atmosphere of the huge building. The malevolent, lethal, ghostly brown mist glided, floated and slithered inside the building, like some evil living entity, patiently hanging around in the atmosphere, awaiting its first victim, or possibly, victims.

The covert nature of the transfer still meant it remained a highly-classified secret. The whole exercise remained undisclosed to the security firm maintaining the site security and the personnel who regularly carried out the routine inspection of the huge facility. Consequently, during one of his routine security inspections, a solitary security guard noticed a wisp of brown vapour, exuding from the base of the huge shutter doors, located at the front of the large building. As far as he knew, the building was completely empty, containing no equipment whatsoever, and entirely safe to enter. Upon opening the large galvanised shutter doors, the lone guard immediately became totally enveloped in the highly dangerous toxic, corrosive vapour, which, like some huge, malicious creature, instantly pounced upon him.

Sadly, due to the amount of IRFNA fumes he ultimately inhaled, the security guard died, his lungs unable to cope with the large volume of the toxic and corrosive chemical which he breathed in. Furthermore, due to its corrosive nature, the vapour also attacked and mutilated a substantial portion of his skin. The circumstances of the guard's grotesque, untimely death instigated a highly secret internal enquiry. Amongst the many conclusions and recommendations arrived at, was the IRFNA should be transferred from the ancient deteriorating aluminium containers, into more substantial and suitable, newly-manufactured larger, stainless steel vessels. There was no way of knowing how many of the old containers were actually on the point of totally rupturing. Thick, high-quality stainless-steel ISO containers capable of holding twenty tonnes of the dangerous chemical, were eventually deemed to be the ideal option, and a remedy to the dilemma.

The MOD commissioned three, '*state of the art,*' stainless steel containers to be custom built, with the scientists, chemists and engineering experts from the research establishment at Porton Down put in charge, to supervise the whole enterprise. The MOD and scientists from Porton Down approached numerous companies with the relevant expertise and knowledge in handling the dangerous chemical. Amongst them, was Chemkiln. The object of the project was to set up and perform the transfer of the IRFNA from the antiquated Russian containers into the new ISO, twenty tonne, stainless steel containers, as efficiently and as safely as possible.

The whole enterprise became extremely complex, necessitating regular meetings with the Chemkiln technical Sales Department, to sort out the nuts, bolts and minutia of the immense task which lay ahead of us.

Finally, following many hours of discussions and high-powered bargaining, our company eventually acquired the contract.

Upon obtaining the final, signed, written agreement, the project was handed over to the General Services department. Fortunately, we had a few months in which to prepare the job, before beginning in earnest. It was decided I would be the team leader. Hence from the outset, I took part in all the meetings. Chemkiln employed a specialised design department which devised and produced the diagrams for the complex equipment required. Working for the MOD, or to be more accurate, Porton Down, we had a substantial budget for the whole venture, some of it paid for in advance. It needed to be, due to the immense cost and complexities of the project.

The official signing of the voluminous contract for the project took place in the first year of the new millennium, with the intention of commencing work in the early autumn of that year. The start date being dependent upon the completion of the fabrication of the new ISO tanks in which the IRFNA would eventually be stored. The manufacturing of the three large twenty tonne tanks proved to be pivotal to the whole project, with the time-line of the undertaking being entirely reliant upon the large tanks being supplied, and in place, before commencement of the work.

Meanwhile, as part of our department's involvement in the project, preparatory work began in earnest. Firstly, a scrubber column unit had to be built, together with an ancillary storage tank for liquid sodium hydroxide, the scrubber unit to contain its own recirculating system. The system inside the medium sized column was designed in such a way, so that the excess fumes of the IRFNA travelled through the column after entering the bottom of the scrubber unit, while the liquid sodium hydroxide flowed downwards, after being pumped up and dispersed through specially designed diffusers, located at the top of the column, giving a counter-current flow of alkaline liquid to the acidic vapour. The inside of the column was also highly packed with Pall rings, helping with the neutralising process, giving good mass transfer. This was the Holy Grail for chemical engineers. The task of the sodium hydroxide was to completely neutralise the acidic corrosive fumes of the IRFNA, before allowing the neutralised, innocuous fumes to vent into the atmosphere.

Secondly, the department purchased a shipping container twenty feet in length, inside which, a complex two-part pumping system was installed. One part of the system was intended to pump the IRFNA into the ISO containers. The other system pumped sodium bicarbonate into the empty aluminium vessel to neutralise any residual dregs of IRFNA left in the vessel, after being emptied. The flows of the liquids into the aluminium containers were manually altered and regulated by a series of interconnecting pipes, pumps and valves.

The operation comprised of various aspects. Each 800 litre, IRFNA container was to be collected one by one using a fork lift truck and then gingerly transported to the area where the container, pumping system, scrubber unit and water bowsers were located. A fair distance from the main office blocks and main entrance of the facility, and certainly well away from the perimeter fence, and out of sight from the prying eyes of any media, as well as being an extremely safe distance from the local community.

During the transport operation using a forklift, from the storage area to the transfer area, and the pumping operation of the IRFNA, the MOD fire brigade shadowed us, in case of any unforeseen emergencies, with their fire tender carrying plenty of water. The crews had been given intensive instructions and lessons concerning the inherent dangers related to IRFNA and how to deal with any incidents. Fortunately, they were experienced in dealing with dangerous materials; after all, the site was an armaments site, used for storing various explosives and ordnance.

The top of each individual 800 litre container had to be unbolted and the top gingerly removed. During the whole operation, the team and I were all completely attired in, hermetically sealed, airtight chemical resistant suits, connected by hose to an air compressor, which fed into the base of an air bottle fitted with a three-way valve. We each carried a bottle on our backs, which was connected to a gas mask. If we had to get away from the container, in the case of emergency, then we could pull one of our arms out from the arms of the suit, put it behind our backs, disconnect the air hose, then switch to the air supply from the air bottle. The bottle contained sufficient compressed air for twenty minutes breathing, allowing us adequate time to get away from the area to safety. It all sounded so simple; what could possibly go wrong?

The first set back was the time-line. As mentioned, we intended beginning the project in early to late autumn of 2000, when generally, the weather would be quite reasonable and not too severe. Unfortunately, due to various reasons, delays occurred in the manufacturing of the twenty tonne ISO containers required for bulk storage of the dangerous material. August became September which itself became October, with still no sign of the project getting underway.

We did not mind the delay, it allowed us plenty of time in which to prepare for the precarious task which lay ahead of us, with all the team receiving plenty of instruction and training in the use of the full, self-contained chemical suits which we had to wear to undertake the project. The team, comprising of Bernard, Ioan, Steve and myself prior to beginning the project, spent hours practicing the escape operation of disconnecting the airline from a compressor and switching to the air bottles on our backs. Manoeuvring our arm out of the sleeves of the chemical suits to perform this difficult operation proved to be no easy task by any means. But it would have to become second nature to us should an emergency arise during the project. Hence, the watchword became practice, practice and even more practice.

I also had to undertake an intensive fork truck driving course in the use and operation of a large cantilever fork truck, I was the only member without an up-to-date fork lift driving licence which I would need to have during the project for Health & Safety and insurance. I had obtained a fork truck licence in the late seventies while employed by another company. That particular licence had long since expired and needed to be renewed for this project.

The delays also gave us time to investigate and consider possible problems. Chemkiln had a maintenance and design department with CAD (Computer Aided Design) capabilities. From this they could ascertain, the 800 litre containers would have to be located on the outside of the shipping container for the pumping of the IRFNA into the large ISO container. There was insufficient head room inside the container to allow removal of the stick of the pump from the aluminium containers. The delay also granted sufficient and adequate time to order and receive the specialised pumps and equipment for the operation. In that respect, the unforeseen delay was a God-send.

It was not until the December of that year before we finally received the go ahead, and a start date for the whole project. Because of the approaching Christmas holidays, the managers at Porton Down decided not to commence work until after the New Year, beginning in the first week of January. We began our preparations in December for getting all necessary equipment shipped to the highly secret ordnance site in Scotland.

Finally, the project began in earnest on Monday eighth of January 2001. Four of us from the department travelled up to Scotland; Bernard, Ioan, Steve and me. That Monday morning, we all met in the early hours at the South Wales incinerator facility, hours before setting off. Unfortunately, the weather did not bode well, it being a bitterly cold, extremely bleak, dark morning. I thought to myself, if it was that cold in South Wales, what would the temperatures be like in Scotland? My apprehension was well-founded. On arrival, that evening in Scotland, the weather was absolutely freezing, and getting colder, with the mercury plummeting all the time. Indubitably, it was not an appropriate time of the year to have a project located outside and open to the elements; especially that far north.

We met scientists and expert explosives personnel assigned with the task of overseeing the project from Porton Down who ensured all safety procedures and agreed methods of operation were fully adhered to. They were an easy-going bunch. The Project Manager and leader Mark Stratford was also laid fairly laid back and a nonchalant sort of character, but extremely mentally sharp and knowledgeable. He knew all the right questions to ask, which meant we had to be well prepared and completely 'on our toes,' ready to answer all of his queries and in-depth interrogation.

Mark fascinated and intrigued me. He appeared to be one of those people who exuded an enigmatic background, fuelling the intrigue and mystery which circulated around him. It quickly became evident, all of his subordinates from Porton Down appeared to know very little about him. Well, at least, that is what they told us. For it may well have been their cover story. Although, I suspect, they were actually telling the truth, and knew very little of his background. Mark seemed to be one of those people who possessed no history, which perversely, probably meant he had quite a lot of history. The guys from Porton Down, told us when it came to explosives and missiles he was extremely knowledgeable. I thought he had something to do with the secret service such as MI5 or MI6 or possibly a '*Big Cheese*,' very high up in the MOD, or perhaps, it was simply my overactive imagination.

For despite his obvious expertise and vast technical knowledge, Mark exuded an unassuming air, a certain laissez-faire, nonchalance, self-deprecation and modesty. He certainly possessed hidden depths, '*Still waters*' and all that. Not once during the project did he ever become flustered, irritated or angry, accepting any setbacks or delays to the project completely in his stride. Which, to me, appeared the major attributes and requirements of a being some kind of 'spook.' But as mentioned, those suspicions may have simply been the machinations of my over fertile imagination. But having said that, Mark's subordinates from Porton Down also seemed to exhibit the inference they were of the same opinion as me. Mark appeared to be some sort of spook. But whatever the true story, there is no denying Mark proved to be a mysterious and enigmatic character, with a poker face, which divulged very little during our time working together. Additionally, Mark would mysteriously disappear for days on end, without a word of explanation to either his subordinates or us.

The first week of the project, we were all shown the site where we were to work and given the necessary documentation to enter the site. For after all, it was a secret MOD establishment and we had all had to be vetted prior to commencing work. The equipment began arriving, the containers, the scrubber column, tanks, fork lift, pumps, stainless steel pipework, chemical suits. The list went on and on. For that first week, our main priority involved setting everything up in the huge site allocated to us, far away from the perimeter and prying eyes. And still the mercury in the thermometer kept plummeting downwards.

On the Monday of the second week, we felt ready to begin our first transfer of the IRFNA. Allocating Steve, the first fork truck driver, to bring the initial tank from the large storage area. We gas tested the building to ensure there were no more tanks leaking inside the building and all the residual IRFNA from the original leaking tank had been completely removed.

Steve gingerly picked the tank with the forks of the truck and began transporting it with the truck to the transfer area, where the rest of us waited patiently. It was quite a spectacle as he carefully moved along the tarmac road in the fork truck, carrying the eight hundred litre tank, with its lethal contents, balanced and securely tied on the forks of his vehicle. Meanwhile, following slowly, a short distance behind, was the site fire engine, containing its fire crew, two of them also wearing full chemical suits. Finally, at the tail of this small convoy, the Chief Fire Officer was driving his brand new, gleaming red Land Rover.

Ioan, Bernard and I waited patiently; all of us knew which tasks to perform, which I had allocated. We had practiced with a completely empty tank to get used to undoing the nuts and bolts, holding the flange in place. After Steve off-loaded the tank, we began undoing the nuts and bolts of the first full tank gingerly placed each one in front of the container. I don't know about the others, but despite the low ambient temperature, I sweated profusely, attired as I was in the full, self-contained, impervious chemical suit, wearing my gas mask connected to the air compressor via the air bottle. I could hear myself inhaling and exhaling through the extremely tight-fitting gas mask. The confines of the mask exacerbating every sound. Nervous energy, my confinement and the exertion of the task causing sweat to effuse out of virtually every pore in my body.

That first transference seemed to take forever. We experienced problems with the compressed air diaphragm pump which continually kept seizing up. All the pumps were completely new, so I assumed, like breaking in a new pair of shoes, it was a combination of this factor, plus the low ambient temperature which exacerbated the extremely annoying problem.

With the first transference under our belt, I decided to call it a day. The first transference of IRFNA had drained us all, both mentally and physically. With it being winter, dusk came early and working at night would be extremely difficult, if not dangerous. The inside of my chemical suit became saturated with my sweat which slowly condensed inside. With the suit being completely impervious, it resulted in my condensed body fluid remaining confined inside the rubber outfit.

After tidying up, we changed into our normal clothes, before heading south to Carlisle, where we had booked our accommodation for the next few weeks. The contract stipulated we should lodge quite a distance from the site eliminating the risk of us becoming friendly with the locals, and inadvertently letting slip our reason for working at the ordnance facility. As well as the issue with the highly dangerous IRFNA, the ordnance site also contained large stockpiles of depleted uranium warheads; a highly emotive and political subject at that time. The MOD had no wish to draw attention to this fact, for fear of opening Pandora's Box, alerting the media and local politicians and possibly stirring up a political hornet's nest, creating a potentially huge PR problem, probably escalating into a crisis, as such things tend to do. No, it was far better to keep the whole enterprise as quiet as possible.

We headed south to Carlisle and our accommodation. Bernard Evans accompanied me in the Rover 75, while Ioan and Steve travelled in the Renault Turbo van. We were all mentally and physically exhausted after that first day, being on the start of our learning curve. After consuming our meal and partaking of a few beers, we all retired early, ready for day two of the project to start in earnest.

The next morning, we had a complete surprise, for, during the night, there had been a severe frost, even more severe than previously experienced. It took ages to defrost our vehicles before heading north on the M6 to the ordnance facility.

Unfortunately, upon our arrival we discovered the inclement weather had frozen all the water lines in the equipment. Prior to donning our suits, we desperately tried defrosting the apparatus, but being located well away from any large sources of heat and hot water, the task proved difficult, if not insurmountable. I took an executive decision and opted to hire a mobile electrical generator together with warm air blowers, to thaw out all the equipment and water lines. Unfortunately, this decision cost us hours, time we could ill afford. The terms of the project committed us to completing the project within the next month which meant a programme of transferring approximately two tanks per day. Already, two days in, the project began falling well behind schedule.

Because of this delay, by the time we eventually thawed out the lines, dusk gradually descended so it was not possible to begin the transfer of any tanks. We called it a day, after ensuring any lines containing water were drained prior to heading to Carlisle and the salubrious warm welcoming environment of our small, cosy hotel. The third day, although cold, we had mentally and physically prepared ourselves for mother nature's inhospitable behaviour towards us. By this stage, we thought the Gods and the elements were ganging up on us, as a sort of paranoia began manifesting itself. After first setting all the equipment up and ensuring all water lines were thawed out and the water running freely, we began adorning our chemical suits. I quickly discovered, inside the wellington boot section of my suit, a substantial amount of ice. I soon realized the ice was in fact the remnants of my sweat, which had made its way down the chemical suit, collecting in my wellingtons. There, because of the low ambient temperature, it slowly agglomerated and solidified.

Before putting on the suit, it became necessary to violently wallop the chemical suit against the walls of the shipping container to break and remove the unwanted, frozen sweat contained inside. It was a task I found obligatory to execute virtually every day of the project, with the negative ambient temperature remaining unremitting. My colleagues also found it necessary to undergo the same ritual, for they too discovered solidified sweat inside their suits. Hence the morning tended to be a cacophony of noise, namely, that of rubberised boots attached to numerous chemical suits being aggressively bashed against metal followed by the tinkling of frozen sweat alighting on the floor of the container. Yet, once again, the start was delayed by having to thaw the lines. Despite having drained the water lines, tiny amounts of obstinate ice somehow managed to remain inside the equipment and lines. This meant having firstly thawing the apparatus out with hot-air. All of these difficulties, slowly and inexorably kept putting the project behind schedule.

That day we managed to transfer only one eight hundred litre container, but the day after that, we began achieving the upward slope of our learning curve, managing to transfer two containers. Things were starting to look up, and we appeared to be getting to grips with the task in hand. That evening in the hotel, the atmosphere was relaxed, and we all had a couple of beers to celebrate. Not too many, I hasten to add, fully cognizant of the work which lay ahead of us the next day.

The day after, it was my turn at driving the cantilever fork truck, transferring the 800-litre container from the large building where they were being stored to the area to perform the transfer operation. Not being an experienced fork truck driver, I must confess to being extremely apprehensive about the task which lay ahead of me. Carrying a corrosive, toxic, lethal cargo inside an ancient, possibly highly corroded container on a bumpy, uneven surface was not my idea of a fun day out, although it certainly got the old adrenaline coursing through my body. The small convoy of vehicles, with me perched on the fork truck, attired in my heavy-duty Tyvek overalls and wearing an ordinary gas mask, with acid filter, headed towards the building which stored the murderous chemical.

Before entering the building, I carried out a gas test to ensure none of the aluminium containers had suddenly split open during the night and confirm no remnants of IRFNA lurked inside. After all, we did not want a repetition of the event from almost a year earlier which had sadly instigated this whole venture. Gingerly, I guided the fork truck up to my designated container. The containers were on metal skids with sections cut out specifically for the forks to go into. However, if my aim was off, then there was always the possibility of the forks missing the skids and piercing the tank above, generating vast toxic clouds of fuming nitric acid. Slowly, I aimed the prongs of the forks into the cut-out sections of the skids. To my utter astonishment, I performed the operation first time, without any mishaps.

Manoeuvring the levers on the fork truck, gingerly lifting the container and its lethal contents a safe, but not too great height, above the ground, after completing this manoeuvre, and securing the vessel to the truck with sturdy ropes, I headed, with my cargo, towards the transfer area, where my colleagues patiently awaited my arrival. My escort and I travelled the kilometre at a sedate speed. From a distance, once again as mentioned earlier, it must have appeared a bizarre scene, the fork truck at the head of the convoy, bouncing gently along, with the fire tender a safe distance behind and at the rear, the Land Rover. My eyes, glued ahead, completely focused upon the ancient container which seemed to bobble, move and jar at every bump or indentation in the tarmac. For those of you addicted to achieving an adrenaline rush, I can highly recommend this activity for achieving such a rush. I speak from experience.

After what seemed like an eternity, the fork truck and its attendant vehicles arrived at the transfer area, where my colleagues waited patiently, ready to perform their part of the operation. Gently setting the container down, I let them perform their specific tasks. This day, I had the easy part, with no manual endeavour, just simply watching my workmates do all the work. I felt no guilt, as we all took turns doing the various aspects of the project. Today it was simply my turn.

After the IRFNA had been removed and the remnants in the aluminium container completely neutralised with sodium bicarbonate solution, I took it to another building where all the containers filled with the bicarbonate solution were being stored. This part of the operation posed no hazard, and it felt almost relaxing driving the fork truck the few hundred metres.

The remainder of the day, I performed the above ritual on a further two more occasions and on both occasions during the transportation of the nitric acid to the operation site, I felt the same nervousness and apprehension accompanied by an adrenaline rush as before.

So far, this turned out to be a record day for the transfer of the IRFNA into the twenty-tonne container. I would like to believe it came about because of my incomparable, expert fork driving skills. However, the truth is my team probably found it easier to do the manual work without me there to hinder them. Now we started 'cooking with gas,' all the planning concerning the project appeared to be coming together. In the words of Colonel John 'Hannibal' Smith from the A Team, 'I love it when a plan comes together.'

For the next few days, we appeared to be getting back on schedule. Unfortunately, after a further three days, one of the diaphragm pumps used for transferring the IRFNA into the twenty-tonne container kept stalling. I made repeated phone calls to our maintenance department back at the plant seeking advice. All they said was to try some lubricant on the piston pushing the bellows of the pump, explaining it was a widespread problem. We persevered for two days and only managed to transfer one tank for each of those days. Once again, we started falling behind schedule. Despite my protestations, no one seemed to heed my concern there was a fault with the pump.

Our new Head of the Department paid a visit to see how the project was progressing. We had a chat and I told him my concerns about the pump. He too agreed with my assessment and insisted the maintenance get a replacement pump and take the one we had back as it was still under guarantee. With two people pushing for this, the Engineering Maintenance department finally relented and contacted the suppliers and demanded we get a new pump. Another delay of a day occurred because of this. We had experienced problems with this pump from day one. When the new replacement pump finally, arrived the difference in performance became immediately evident. There had indeed been a problem with the original pump. After receiving the new pump, we started transferring three tanks a day. It looked as if we were sure to maintain the schedule. But fate certainly has a way of spoiling things.

The final major problem which occurred during the project occurred almost half way through. And it gave me quite a fright I can tell you. We determined when the containers had been completely emptied of the IRFNA, by weight. Before we began the project, the tare weights of a few old empty containers were taken to determine an average for the weight of each aluminium vessel when empty. During the transfer operation, the full containers were placed onto beam scales capable of taking weights of up to three metric tonnes, with a digital readout. This day, for some inexplicable reason the beam scales gave erroneous and misleading readings, which indicated all the IRFNA had been transferred. We closed valves and opened others allowing transfer of the sodium bicarbonate into the presumed empty aluminium container, to neutralise the assumed small amount of acid remaining at the bottom. The small volume of generated fumes vented through the scrubber column and all would be well with the world.

Unfortunately, instead of being empty, unbeknownst to us, almost two hundred kilogrammes of concentrated IRFNA remained in the aluminium vessel. We began pumping the bicarbonate solution into the vessel which began immediately reacting violently with the acid. The plate containing the down pipe through which the bicarbonate was entering the tank had only been hand tightened with the eight nuts and bolts.

Suddenly, the liquid inside the container began to effervesce and exude between the plate and flange at the top of the vessel. Brown, frothy liquid began exuding out of the vessel and running down the sides. Masses of brown fumes also started to be generated. I immediately gave the order to stop pumping the bicarbonate into the vessel and without thinking grabbed two spanners and began tightening the nuts and bolts holding the plate and flange together to try and plug the leaks, using all of my meagre strength to tighten them down.

My loyal team, heroic to the last, stood back, remaining at quite a safe distance, while I endeavoured to stem the flow of reacting liquid from the top of the vessel. The guys from Porton Down began shouting from outside the shipping container informing me the scrubber column could not cope with the amount of the brown fumes being fed into it, with brown clouds of IRFNA venting to the atmosphere. Fortunately, we were located so far from the perimeter and at the back of the facility that the brown fumes could not be seen from the main highway and most importantly, a safe distance from the civilian population and completely out of harm's way. Thankfully, being Scotland, with its heavy winds, the fumes dissipated very quickly.

Meanwhile, inside the confines of my chemical suit, the sweat poured out of me, and I found difficulty breathing through my gas mask, all the exertion causing me to breathe faster, as I attempted to get more oxygen into my lungs. Not only that, my chemical suit visor and gas mask began to mist up, affecting my vision. Finally, I managed to tighten all the nuts and bolts sufficiently to stem the leaks. The reaction, taking place inside the aluminium container caused it to vibrate and become quite warm because of the violent, exothermic chemical reaction.

Gradually, the reaction inside the aluminium tank subsided, as all the remaining bicarbonate finally reacted with the acid. Consequently, the fumes being vented at the top of the scrubber column, also slowly abated. We waited a while, allowing the vessel to cool and finally uncoupled the plate on the flange, replacing the lid. The aluminium container was later moved back to the facility where all the other vessels holding the IRFNA were being stored. The solution in the vessel had now become contaminated and unsuitable to be used, we later neutralised the remaining solution with the more alkaline caustic soda under strict controlled conditions.

Follow that excitement, I felt completely exhausted the incident had been physically and mentally draining. It reminded me of the incident while working for West Mercian Oil and the desorption column, with the hot mineral oil cascading over me.

That was the last major incident and we purchased a new beam scales and made certain it was thoroughly checked before starting. As the days progressed we started arriving at the top of the learning curve finally ending up transferring a record six containers in a day and the project was eventually completed two days ahead of schedule, much to Mark's gratification and relief.

A few weeks later, British cruise missiles with conventional warheads began flying towards Iraq after Saddam Hussein antagonised the western governments. Some no doubt propelled by the IRFNA, we recently transferred. No wonder there had been such a rush and panic in the final weeks to complete the project.

During the project, I handed in my resignation. Bridget had been demanding and pleading with me to finish with Chemkiln, because she hated me being away for weeks at a time. Because I loved my new partner so much, and desperately wanted our relationship to succeed, I eventually agreed. With great reluctance, I handed in my notice.

During my last month, I had a long chat with the

Chemkiln MD. He, together with other directors had received a complimentary letter from Mark in which he highly praised the whole department, me included, saturating the letter with superlatives and hyperbole. It was mentioned in despatches by a prestigious government department. High praise indeed. The MD also told me he was sorry to see me go but at least I was finishing on a high note with no hard feelings from either party. He knew of my personal reasons for terminating my employment with the company, as did everyone in the department.

A couple of weeks later, I left the site for the last time, filled with a mixture of sadness and trepidation. I loved my time at the company, it had been very good to me, providing me with a lot of memorable experiences and adventures and good pay. Now I was forsaking it for another. The things we do for love eh?

Unfortunately for me, that love would later prove to be totally misplaced.

CHAPTER 19

Shortly after taking the momentous, and let's be honest, my ludicrously stupid decision to quit Chemkiln, there followed a period of uncertainty about my future, while I desperately sought employment; sending off many copies of my CV to a multitude of local companies. In the June of that year, a letter dropped on the door mat of my new home on the outskirts of Llanelli, where I now lived with Bridget and her two young sons. The correspondence came from a business called Soleil de Mer, which I discovered, sold cosmetics, toiletries and face packs. I must admit, not one of my specialised hobbies or subjects. The contents of the letter requested I attend an interview for a future, unspecified position within the company, based on the outskirts of Swansea. After scrutinising the letter, I decided to reply, informing the company I would indeed be attending the interview scheduled for a week or so later.

I arrived early for the interview held in the offices at the company's warehouse facility in Swansea, presided over by the Manufacturing Director, Harry Chivers. Harry turned out to be a man in his late fifties, of average height and build, with thinning grey hair. During the interview, he informed me the contents of my CV appealed to him very much, particularly my experience involving chemical processes and troubleshooting manufacturing problems. He further elaborated, by explaining he wanted to create a new position within the company, intending to incorporate the new role of a process engineering trouble-shooter within, and part of, the Quality Manager's duties.

However, he had a dilemma, the company already had a Quality Manager, but Harry wanted the incumbent to take on another new position, that of IT Manager, a position he felt he would be far better suited to and more in tune with. Harry had another problem, convincing his fellow directors to go along with his wishes. The additional new IT position would mean increasing the wages bill. Generally, the company held board meetings every month. Unfortunately, there would not be another executive meeting for a few weeks. Harry stated he wanted me working for the company as soon as possible. He had another concern, the possibility of me accepting a position with another employer. Although, he had nothing to worry about on that score, as I had only received loads of rejection letters. But I wasn't going to tell him that.

After carefully deliberating the problem, he arrived at what he deemed to be a compromise, and a feasible solution in resolving his dilemma. Harry wanted the Materials Manager to take on an assistant to help take some of her workload. This new job position had already been agreed to and signed off by the board. Harry suggested I become her assistant for a temporary period until he could get me into the Quality Manager's position. I had been in the same analogous situation years earlier, when I joined Birchwater in the seventies as a Trainee Manager. It transpired not one of the Management Trainees ever became a fully-fledged Manager. Ultimately, this was one of the reasons, along with some other personal issues, why I eventually left that company, hence my reluctance in accepting the temporary position now on offer. If Harry's plans did not come to fruition, I would be stuck in the job as an assistant. Harry told me to go away and think about his proposition.

After careful deliberation and discussing it with Bridget, I decided to accept Harry's offer. Well, at least to have a meeting with the Materials Manager. By this time, following my drastic and foolish step to leave my lucrative job with Chemkiln, my savings had dwindled considerably. Accepting the tenuous position would at least give me an income, however small. After phoning

Harry, informing him of my decision to accept his proposition, he arranged a meeting with Ava Morgan, the Materials Manager. The objective of the meeting was to determine whether she would be happy for me to work under her.

The day of this second interview, meeting, call it what you will, arrived. I must admit to feeling ambivalent about the whole situation, wanting the job for some income, but on the other hand, not really desiring it, hoping something more rewarding and appealing might come along. The interview went okay, not fantastic, just okay. Ava turned out to be a thin, hard-faced individual in her mid-forties, seeming to possess very little warmth or sense of humour, and appearing to harbour masses of insecurities. Most of the interview involved her telling me about her job helping maintain stocks of materials supplied by the manufacturers. Basically, informing me how well she did it, not really seeing the point of employing anyone to assist her, only going along with the situation in order to placate her boss.

Soleil De Mer did not actually manufacture their products, instead choosing to outsource the manufacturing operation to suppliers. First in the supply chain came the bulk suppliers of the products, second came the laminate contractors who manufactured rolls of metal foil, then finally, the fillers, who put everything together; putting the bulk soaps, face packs, foot scrubs, or exfoliates into sachets or bottles. During my interview with Ava, she kept reiterating and emphasising the fact she believed I was far too qualified for the position. Evidently hoping I would not take it.

Following my meeting with Ava, Harry Chivers and I had another one to one chat. He still offered me the position, obviously instructing Ava that she had to take me on as her assistant, with no objections permitted on her part. During our discussion, he kept stressing he wanted me to accept the position temporarily, until he could get me in as the QA Manager. Having had time to consider the position, I agreed to take the job, commencing the following week, but with huge reservations on my part. Unfortunately, my initial assessment of Ava had not been misplaced, most definitely not my notion of an ideal Manager or person to work under, for she often took great delight in pointing out any gaffes on my part, or anyone else for that matter, and immediately informing Harry Chivers of any blunders by any one of her colleagues.

Every cloud has a positive aspect, because of Ava's petty vindictiveness, I was determined to get to grips with the Assistant's job as quickly as possible, learning the stock control and accounting computing system called Sage, together with the company paperwork systems. It appeared wise not to make too many mistakes, for Ava made it obvious to everyone in the open plan office when one had been perpetrated. So, that became my motivation and focus in getting to grips with the office systems as quickly as possible. Let's say Ava caused me to have an extremely steep learning curve. My line manager also tended to be duplicitous, one minute talking nice and friendly to fellow colleagues, the next vociferously criticising them behind their backs. I frequently wondered what she said about me when out of earshot. Nothing good, I'll be bound.

I had only been working in the open plan office a few weeks when an event occurred which affected the start of the new millennium, with consequences for the decades to come. I refer of course to what is now known as 9/11. It is one of those historic events when everyone remembers where they were, when they heard about it, who told them, what they were doing at the time etc. William Brewer, the Product Development

Manager, mentioned it to me during the lunchtime. He told me a small light aircraft had crashed into one of the twin towers at the World Trade Centre in New York. I suspected the light aircraft had crashed because the pilot had had a heart attack or some other form of incapacitating seizure. Someone in the office turned on a small radio. Gradually, as the horrific events unfolded, it became obvious something more sinister and gruesome appeared to be taking place. That afternoon, we all listened intently to the radio, with not much work getting done. Increasingly, our small office, along with the rest of the world, became aware of what was unfolding.

It is fascinating how globalised and interconnected the world is. One of the customers in America informed us his wife worked in the personnel department of one of the airlines, and how she had to console the relatives of people working for the company, and who had the misfortune to get caught up in the horrific events that day, eventually paying with their lives, his wife having to contact the relatives, imparting the tragic news. Yes, it goes to show how interrelated and connected the world actually is.

True to his word and to my immense relief, after a couple of months, Harry Chivers managed to obtain his political wishes. Finally persuading the board, along with the company owner, Dean Fairfax, to create a new position of IT Manager, to which he re-assigned the incumbent QA manager. The vacant QA position he then allocated to me, with additional technical process responsibilities, assisting the manufacturers with any technical problems when producing Soleil De Mer products. I also had to establish a small laboratory, checking the bulk products for such things as viscosity and pH, basically, checking up on the bulk suppliers. Additionally, I was also given responsibility for being caretaker for all the formulations concerning all the products. The QA section was being improved and uprated. It became my task to implement those changes. My first few months in the job passed quickly. I tended to be extremely busy, to say the least, trying to get to grips with the many products on offer by the company, as well as purchasing analytical equipment and instructing the team working under me how to use it.

During those first months in the job, I eventually met Dean Fairfax, the MD and owner of Soleil De Mer. The personnel working for Soleil De Mer, gave the company the abbreviation SDM, the French proving far too difficult to pronounce for most people. Dean normally based himself in the head office located on the outskirts of Guildford, from where he disseminated his authority and control of the company. The directors generally did the day to day running, but SDM ultimately remained Dean's company, his conception and his baby. It was he who gave the company its fairly pretentious French name. Dean proved to be your stereotypical, *rags to riches* story; He began by making soaps, cosmetics and face packs in a small shed located at the back of his house, touting for business by attending cosmetic trade fairs. To save on expensive outlays such as hotels and their overpriced meals, Dean often slept in the back of his dilapidated, old white van, surviving on home-made vegetarian meals. By the time I joined SDM, it had become a well-known and well-established multinational company, propelling Dean into the multi-millionaire income bracket.

I had not been forewarned about Dean, who turned out to be quite an eccentric, unpredictable individual. A strict vegetarian, who, during his younger years was even appalled at the thought of wearing leather shoes and clothes. Instead, choosing fabrics not originating from the skins of dead animals, but wearing cloth sandals on his feet instead. A veritable Hippy. Nothing wrong with that, after all, many moons before this period in my life, I too had once been a long-haired Hippy, during my halcyon, carefree days of being an undergraduate at University. Dean also refused to own any cars upholstered with leather.

The first time I encountered the eccentric Dean, I should have been wearing highly tinted sunglasses. For, despite his elevated position within the company, Dean had no dress sense whatsoever. He turned out to be a man of average height and build, in his mid-forties, with a mop of thick, grey hair, his attire mostly comprising of excessively bright clothing, with an eclectic mix of highly contrasting colours. He turned out to be the Willy Wonka of the cosmetics industry. During my first encounter with our leader, he was wearing a dazzlingly bright lemon jacket, accompanied by a vivid green shirt, pale blue trousers; he also sported a multi-coloured, spotted, lilac tie. All this supplemented by extremely vivid pink canvas shoes. It only required a large, coloured top hat to complete the scene. You could virtually observe him approaching from a fair distance away, like the Coca Cola truck, but without the music. I quickly learned this turned out to be his normal form of attire, helping him to still maintain some form of tenuous allegiance to the psychedelic colours from the Hippy period in his life.

Notwithstanding his attire and his idiosyncrasies, Dean proved to be an affable, friendly individual, quite garrulous and chatty. Unfortunately, during my time employed by SDM, he would often cause me, along with many others within the company, no end of problems and headaches with his strange, idiosyncratic wishes and behaviour when dealing with the company's products, often exhibiting extremely specific, weird, curious requirements and desires.

Not only Dean, but SDM as an entity, turned out to be a strange and bizarre company to work for. Nevertheless, it was extremely interesting. Harry

Chivers also turned out to be intriguing. He kept imparting stories of his time spent in the army, where he apparently achieved the rank of Captain. Yet he never used the epithet of that rank, which I considered strange, being under the distinct impression anyone with an officer's rank always used the rank before their names. He frequently told people, how he served in Northern Ireland and at one stage, during his allegedly illustrious career, how he once held responsibility for the movement of Britain's nuclear arsenal around Britain.

Sometimes scepticism is a reasonable trait to possess. When I met Mark from Porton Down, it became obvious he had something about him, a certain *je ne sais quoi*. His deportment, together with the way he carried himself and his general behaviour, indicating he had experience, and knowledge. He exuded obvious immense leadership qualities. Somehow, Harry Chivers just did not seem to convey those same qualities. I certainly could not envisage the MOD leaving the responsibility of moving Britain's nuclear weapons around the country to a lowly, newly promoted Captain. That responsibility, I would have thought, should rest on the broad shoulders of, at least, a Colonel or Brigadier.

The first few months of my time spent at the cosmetics company, proved to be quite enjoyable and interesting, visiting the companies who made or packed the products, which SDM then sold. Basically, as previously mentioned, the suppliers could be split up into three major categories in the chain of manufacture.

Of course, there was the usual, obligatory political intrigue within the company, with the directors and managers vying for a superior position in the hierarchy, mainly by gaining Dean's approval, praise and favour. I had worked in enough places to observe these shenanigans and tried to stay away from all of the jockeying for position and status within the company. For example, there was William Brewer, the Product Development Manager, who continually travelled to Horsham for meetings with Dean. I learned William had originally been the Plant Manager at the old, former, smaller warehouse, located a couple of miles away. He had employed Harry Chivers to work under him. Unfortunately for William, Harry turned out to be an ambitious and scheming individual. Harry became involved in the planning of the relocation to the premises we now inhabited.

The planning of the move meant Harry met Dean on a regular basis. Quite suddenly, William discovered he now worked under Harry, instead of vice versa. Adding salt to the wound, Harry became elevated to the position of a Director on the Board. Consequently, by the time I joined the company, William festered with resentment and discontent, taking any opportunity he could to denigrate his boss. Ah, more office intrigue. I cannot remember the amount of times I observed company politics and self-advancement from a distance with other businesses. Even places like the esteemed White House are not immune to such political hiatus. SDM tended to mirror that establishment inhabited by an unpredictable, slightly deranged leader, with a chaotic style of management. In his defence, notwithstanding his unpredictability, Dean did retain a certain personality and charm.

My first problem brought on by Dean's idiosyncrasies, occurred within a few months of becoming the QA Manager. As part of my duties, I had to deal with customer complaints. I began noticing a dramatic rise in complaints linked to one of the more popular products, The Dead Sea Face Tonic. The 100ml sachet comprised of a fabric mask immersed in a blue solution. Ostensibly, the blue solution represented the Dead Sea. I feel it my duty to point out to the reader, certain basic facts. The Dead Sea is not actually vibrant blue, and nothing like the solution contained in the SDM sachet of that name. Instead, the actual Dead Sea is a distinctly muddy brown colour, and extremely dirty.

A brief period before I joined the company, Dean insisted the solution contained in the Dead Sea Face Tonic sachet, become an even more vibrant blue. Dean became adamant, insisting upon his wishes being carried out, despite protestations from his Formulation Chemist and other Directors. Being the outright owner and answerable to nobody, Dean had his wish, and the blue colour in the formulation became enhanced by more than fifty per cent. Suddenly, one could determine the customers who purchased the Dead Sea Face Tonic supplied by SDM, due mainly, to the distinctive, pale blue discolouration they exhibited on their faces, as they all took on the appearance of Captain Doreen Lewis in Private Benjamin, following her shower, after the platoon filled the sprinkler head with blue dye, and she ended up resembling a cadaver or zombie.

Because part of my duties included reviewing customer complaints, suddenly it became evident the amount of complaints relating to the Dead Sea Face Tonic started to increase exponentially. This was because gradually, more and more faces of SDM customers using the Dead Sea Face Tonic began assuming the distinctive blue hue. When interrogated by Dean as to the major product issues, I had no hesitation informing him of the problems concerning the Dead Sea Face Tonic.

I suspect Dean did not believe me, or at least, thought I was exaggerating. When the directors also began telling him about the customer complaints and the fact they were becoming vociferous, with threats of litigation in the offing, he quickly began taking notice. Dean resolved the problem by replacing the blue solution with a blue cloth, containing a permanent blue colour, which would not discolour the faces of the customers. The actual solution contained in the sachet became colourless, with the entire old product ultimately removed from the outlet shelves.

Dean had an accomplice when it came to causing problems for the management team and the workforce. For Harry Chivers also proved to be no slouch in the department of generating problems and difficulties for others to resolve.

During my initial period working for SDM, while walking along the walkways of the warehouse, I began observing enormous amounts of distorted cardboard boxes which contained sachets. The boxes ballooned and bulged out quite markedly. Exactly the same problem one experiences when putting on clothes after suffering weight gain. I thought no more about it, until one day, when Harry called me into his office. He began explaining the problem in great detail, together with his concerns. The cardboard boxes, obviously under so much strain, contained sachets of a face product called Milk and Oatmeal, basically porridge. Why anyone would wish to plaster their face with porridge is beyond me, but as the saying goes up north, 'nowt so strange as folk!'

Some months earlier, during the manufacturing process, some bacteria called lacto bacillus had somehow managed to contaminate the product. The lacto bacillus reacted with the oatmeal, generating carbon dioxide. The sachets, were in fact, slowly fermenting inside. Unfortunately, the carbon dioxide had nowhere to go, so the pressure built up inside the sachets causing each one to balloon out. When you have a hundred ballooning sachets packed into a cardboard box, the box inevitably also starts to swell out.

Harry exhibited great reluctance in scrapping these

old sachets, most probably because of Dean. There were almost 200,000 of them in total. Harry came up with a possible solution. SDM employed people who worked in a cabin putting labels onto the sachets, changing languages for the different countries. SDM considered it a cheaper option to stick labels on the sachets for the individual countries, rather than have multiple sachets with different languages printed on them. When the company received an order, from say Poland, labels would be printed in Polish and stuck over the English print. Harry suggested the personnel in the cabin, could be utilized in solving his Milk and Oatmeal problem. They could stick a sterilized pin in each sachet, and after allowing the carbon dioxide to escape, assisted by gently pressing on the sachet, they would then place a small round blank label to cover the hole they had made. He estimated each operator would take no more than twenty to thirty seconds to complete the whole operation. With ten operators working seven hours a day, Harry calculated the entire stock of Milk and Oatmeal could be rectified within three weeks.

I discovered very quickly, how Harry found great difficulty in accepting any criticism, or dissent. But somehow, he had to be dissuaded from instigating this hair-brained scheme of his. I believed, if it could be effectively demonstrated his plan would just not work in the practicable sense, he could be manipulated to abandon the whole idea. Adopting psychology to persuade him to relent with his concept, but not making the subterfuge appear blatantly obvious, Harry would have to be surreptitiously cajoled and manipulated. I acted completely naïve and innocent, requesting he demonstrate the operation as if I did not understand or comprehend the idea. The other managers and supervisors egged me on. They had no desire to implement the operation. For, they knew full well the pitfalls, and how exceedingly difficult it would be to carry out.

After my discussion with Harry, he suggested I observe a demonstration of the operation which he would perform himself, while I timed him using a stopwatch. I collected everything required for the operation, an extremely distorted box of Milk and Oatmeal sachets, some sharp pins, a small container of isopropyl alcohol, necessary for sterilizing the pin, a reel of small label dots, a stopwatch and loads of absorbent tissues, which I felt would be definitely needed.

I entered Harry's office, carrying all the accoutrements, necessary for the exercise, contained in a large box. My boss took on his paternalistic attitude, as if talking to a child. In his patronising way, he described the process, grabbing hold of a Milk and Oatmeal sachet. He instructed me to start the stop watch. First, he stuck a pin in the sachet and slowly compressed the sachet. Suddenly, the contents began exuding all over his highly polished, mahogany desk. Harry desperately tried stemming the leak, which only made matters worse. He then tried cleaning around the hole to stick on the small label, but due to the porridge contamination, it refused to adhere. He then took off the small label, but once again, the contents leaked out. My stopwatch relentlessly ticked away. One minute thirty-five seconds passed by, and Harry had still not completed one sachet. After three minutes, he stopped and said he would have another go at it. After five attempts, all without success, Harry finally admitted defeat. I pointed out he expected the operators in the cabin to do this easily, in twenty seconds and for thousands of sachets.

'Mmm,' he retorted despondently, 'perhaps it is not such a clever idea after all.' He had finally come around to my, as well as others, way of thinking. We had not even argued: the power of diplomatic, manipulative persuasion. You just cannot beat it.

Another instance where Harry instigated a horrendous blooper, involved the setting up of an additional overspill warehouse to accommodate the additional items being produced.

The company began to expand. Primarily, because of Dean and his ideas, an explosion in new products came about. With all this diversification, the company experienced a distinct lack of storage space. The directors decided to lease new premises. The board, which, to all intents and purpose, meant Harry, decided upon leasing an old deserted factory as an overspill site, located a few miles away, at a place called Fforestfach in Swansea. Harry had some of the warehouse staff commence installing shelving, bolting the units to the existing concrete floor.

Harry had little or no regard for health and safety. The warehouse staff worked in the premises without electricity, water or proper canteen facilities. The owner of the site, as far as I ascertained from talking to the other managers, appeared to be a bit of a wide boy, and fairly devious. But it seemed, Harry had been completely taken in by him. One day, Gerald, one of the supervisors came into the office, complaining vociferously about the distinct lack of amenities on the site, and how his team of men had no power tools to work with because there was no electricity supply.

'Of course, there is electricity, the owner had his men using power tools. I saw them using electrical power tools myself.' Harry answered aggressively, taking an extremely belligerent tone.

'That's because they had an electrical generator.' replied Gerald, who then glared at Harry, before walking away.

I observed the expression on Harry's face, open mouthed, that of a person in complete shock, and at a loss for words, like a landed fish gulping on the shore.

Not typical Harry Chivers.

Upon further investigation, Harry discovered the main electricity cable, supplying the site, had been disconnected quite a distance from the premises, when the building had been decommissioned years earlier. It would cost tens of thousands of pounds to get the supply reconnected. The owner of the building refused to pay for it. So that was the end of that. The project ended up being ultimately shelved... no pun intended. Yet, Harry received no reprimand for the time, effort and money spent on the whole enterprise.

Harry had once been responsible for moving Britain's nuclear arsenal around the country? I think not!

Mainly because of Dean, it became necessary to invest in expensive equipment to aid the fillers, due to his bizarre requirements and demands. After a period of time, I began to see why Harry required the services of a chemical engineer, because of Dean's weird requirements. The owner often demanded particular desires, especially when it came to the formulation and production of a material called, Men's Dead Sea Mud Mask. Dean, being his usual idiosyncratic self, stipulated the formulation should not contain any homogenisers or emulsifiers, to aid keeping the solids in suspension. Our illustrious leader professed to be an environmentalist, yet I could not see anything remotely environmentally friendly in digging up areas of the Dead Sea to produce material for plastering on people's faces. All in the name of vanity and narcissism.

The company which put the product into the sachets had tremendous problems, because of Dean's demands; I had to assist them in fulfilling their contractual obligations, which they found difficulty in keeping, particularly with regards to the Men's Dead Sea Mud Mask. The producers of the bulk were located in St. Leonards near Hastings and the fillers in Telford

Shropshire. The bulk manufactures shipped the product in pallecons. Basically flat-packed boxes erected to contain a plastic bag inside and the bulk material pumped into the plastic bag, weighing about one metric tonne. Because of Dean's quirky demands, specifically, that prohibited the use of homogenisers and emulsifiers, by the time the materials journeyed one hundred and fifty miles, experiencing a large amount of bouncing and jarring, the solids inevitably settled at the bottom of the plastic bag contained within the box.

By the time the product reached the filler, the solids of the product remained like concrete at the bottom of the pallecon, with a layer of water on top. The pumping equipment normally used by the filler, was unable to cope and pump the solidified material. I had to purchase rather expensive stainless-steel reciprocating pumps, with the capability of transferring the material to the filling machines, the company also had to keep the material continually recirculating, in an effort to keep the solids in suspension. All this because of one person's idiosyncratic desires and wishes.

This was the company I now worked for. Oh, deep joy.

CHAPTER 20

The company, when I say the company, I basically mean Harry Chivers, required I travel the country, visiting the suppliers helping them get to grips and resolve any production or product issues. Or give them help when manufacturing a product for the first time. Just like my time spent working for Chemkiln, I enjoyed travelling the country, meeting the different suppliers. The task not onerous or burdensome. I appeared to be getting on well within the company, that is until one fateful night.

Mike, one of the supervisors working in the cabin, invited his workmates and colleagues to his engagement party being held in one of the local clubs, which boasted a huge function suite capable of taking a few hundred people. Mike not only invited his colleagues, but their families as well, including Bridget and myself. Bridget and I had not been out socially for a while, and I jumped at the chance. It also meant Bridget and her two boys would also be able to attend without worrying about finding a babysitter.

The evening of the party arrived, and I introduced Bridget to Harry Chivers, who also brought his wife, Cynthia along. We all sat down together on one of the tables. During the evening, I chatted to Cynthia, a nice personable lady, while Harry appeared to be getting on swimmingly with my wife as they conversed incessantly. I thought nothing of it, as they appeared to be deep in enjoyable, friendly discussion.

The following week Harry called me into his office and told me he wanted to have a quiet chat. He let me know what he and Bridget had been talking about that evening of the engagement party. Apparently, Bridget had told him, in no uncertain terms, she did not like me working away and needed me at home while she attended evening classes a couple of nights a week in order to further her career. She needed me home on those evenings to look after her two boys. The increasing, irregular frequency of my time spent away caused problems, and she requested, no rather, she demanded, Harry no longer send me away to visit the suppliers.

Now Harry had misogynistic tendencies, and did not like being dictated to, particularly by a woman. My standing with him became diminished, because I appeared to be ruled by my wife. He took me on specifically to meet regularly with the suppliers, now all that seemed in jeopardy. He told me he did not like the situation and left it at that. Consequently, my position within the company began to alter, I began experiencing the feelings of being shunned, and not involved to the extent I had been before. Harry's amiable persona towards me altered drastically, and he barely said hello to me during the working days. He obviously now considered me to be unsuitable management material and now *persona non grata*. Unfortunately, this state of affairs continued for quite a few months. Bridget had caused me to finish with Chemkiln, now she seemed intent upon ruining my prospects with SDM.

As well as having grief at work, I also had a tough time at home, with Bridget tending to become aloof and extremely cold towards me and frequently argued about what transpired that evening with Harry. Eventually, one day she informed me she longer loved me and confessed to having an affair with her best friend, Pamela's husband. Telling me she loved him very much. Immediately, everything became clear, and I had a *light bulb* moment. What a complete and utter Dumbo I turned out to be, Bridget didn't need me to at home childminding for her to do evening classes or her charity work, for animal rescue. She needed me at home to enable her to visit and spend time with her lover, who had recently left his wife, and supposedly Bridget's best friend.

Bridget lied to her friend as well as me, telling Pamela she had not seen or heard from her estranged husband in ages, when in fact they had been seeing each other on a regular basis. Not only that, I later discovered the affair had been going on for quite a while, even before Bridget and I knew each other. Now that he had finally left his wife, he wanted Bridget to move in with him.

I visited Bridget's betrayed female friend who opened up to me, informing me about Bridget's colourful sexual past. It turned out my new wife had quite a reputation as a man eater and was responsible for quite a few marriage breakups. Invariably, with Bridget tending to be the guilty party in those situations. On one occasion just after we were married she was chased around the shopping aisles of ASDA by one of the angry wives who threatened to *knock her teeth* out if she ever caught up with her. Fortunately for Bridget and her teeth, she managed to escape. There was another story I discovered afterwards. I was not even her third husband, but her fourth, for God's sake. After talking with Pamela, it became evident I did not know my wife at all, she was a complete stranger.

Shortly after all these revelations, I left the marital house. Bridget should have been christened Jezebel or Messalina. She had completely ruined my life, along with countless others. First, persuading me to leave a well-paid job which I enjoyed, now ruining my future prospects with my new employer. I was in hell, this woman was decimating my life, which was now inexorably falling apart.

Some of my colleagues in SDM discovering my predicament, realizing my somewhat pitiful, mental state, found me accommodation with one of the guys who supervised the night shift in the cabin and who rented rooms in the large house on which he had a large mortgage. The move became easier with his house being situated in Swansea and much closer to work. My new landlord turned out to be an amiable German named Jeurgen. Germans have a reputation for not possessing a sense of humour. I don't know if this belief is true or not, but Jeurgen certainly dispelled it, possessing a wonderful sense of humour and amiable disposition.

Possibly the exception which proves the rule, but his sense of humour helped with the transition. Jeurgen frequently joined me for a few drinks in the local pub which helped alleviate my situation by drowning my sorrows. The house he owned was situated near Swansea City centre and closer to SDM with the additional bonus of making it easier with less travelling time. Within a brief time, Bridget filed for divorce citing my unreasonable behaviour as being the grounds, filing before I could put in my divorce citing her for adultery. It is indeed an ill-wind which blows nobody any good. Following my split from my new, short-term wife, life improved dramatically. I began to socialise more with my colleagues from SDM and going out more.

Working away proved not to be the hindrance it had been while living with Bridget. Harry Chivers became aware of my personal circumstances and after a discussion, I told him I could travel more. So, the situation began to improve on the work front as well, and I began to look much healthier and happier. The few months of living with Bridget following her revelation about the affair had been a mental strain, with Bridget constantly picking arguments, obviously in an attempt to speed up my departure.

As mentioned, every cloud has a silver lining and positive aspect. During this period, I befriended a widow named Christine who worked at SDM. During any works functions we often gravitated towards each other. We talked a lot and during one of our discussions informed me she had originally lived in Mumbles, a suburb of Swansea. During her teens she had a teenage obsession with horse riding, often taking customers of the stables out on pony treks. I told Christine, I had once gone pony trekking in Mumbles and fell off my horse after the young girl in charge of the ride decided to take her leading horse for a gallop.

Of course, all the other horses immediately followed suit, mine included. Unfortunately for me, the horse took it into its head to hurdle a fairly large log, resulting in me falling headlong into the grass. Luckily, I only sustained a few minor cuts and bruises, along with a badly damaged pride and ego. When I told Christine about my experience all those years previous, she began smiling. When I asked her why she found the story so amusing, she replied, 'I was that young girl.'

We even talked of how we both frequented a discotheque in Mumbles where the DJ loved playing Creedance Clearwater revival. Although we must have been there at the same time, neither of us recollected the other

It certainly is a small world, although I wouldn't like to have to paint it.

We began seeing more of each other but kept our passionate relationship a secret. Firstly, I did not want Bridget finding, there was no way of knowing what she would do, being an unpredictable and volatile individual. She may not have wanted me but could make it difficult for any female showing an interest in me. Secondly, Christine had adult children and she did not know how they would react to the fact their mother was seeing another man after becoming a widow. So, we thought it wise to see how things developed before informing the world of our blossoming relationship.

However, there was one person who knew of our

developing relationship. Christine's close friend, Debra, who kept quiet about the whole situation. The first Christmas after we started seeing each other, SDM had a Christmas dinner and dance in one of the large hotels in Swansea. Christine and I decided to make our own way there. She travelled by taxi with Debra. I travelled alone from Jeurgen's house, also by taxi. We arrived at the luxurious hotel at almost the same time. On seeing me, Christine introduced Debra, but only after first introducing her to Dean Fairfax and Harry Chivers, even though Debra and I had known each other for months.

Christine began, 'Debra, I would like to introduce you to our Quality Manager, Vinson Chard.'

Now, I don't know why she did it, probably through nerves, and concerned she might give our little secret away? But while I shook her outstretched hand, Debra immediately curtsied. Not a small curtsy but an extremely deep, supplicant curtsy, while at the same bowing her head, time quietly muttering,

'I'm deeply honoured to meet you, Mr Chard,' addressing me as if I was the Pope or part of the higher echelons of Royalty. I went to pull her up, but in the process, of bending forwards to perform the gentlemanly action, I spilled most of the contents of my full, cold pint of beer down her generous cleavage. Pandemonium then ensued as Debra emitted a high screech, while at the same time leaping to her feet, her cleavage, sodden, and dripping with ice cold beer. Why did Debra curtsy? God only knows, she didn't even curtsy when introduced to Dean Fairfax, the MD, a person who completely outranked me in the scheme of things. I tried to apologise while, everyone gathered around to see what the commotion was about. I also tried wiping Debra down with the nearest available napkin, but she angrily pushed me away, while everyone else kept telling me what a clumsy clot I was.

Meanwhile, Christine hid her face in a mixture of embarrassment, shame and utter disbelief at the risible scene slowly unfolding before her very eyes, and secretly wishing the ground would open and completely swallow her up.

Now as I recall that incident some fifteen years on since that night, Christine and I are now happily married. Debra and I are still friends and we often reminisce and have a laugh about that evening and the cold beer being thrown down her front.

I must thank Bridget, who turned out to be my transient relationship. She enticed me from Stella and through her enticement, I met Christine. We have now been happily married for twelve years and is one of the best decisions in my life. As I said earlier, it is indeed an ill-wind which blows nobody any good.

CHAPTER 21

During my travels working for SDM, I became friendly with many of the suppliers, building up a good personal and business rapport. Whether their bonhomie was purely an act on their part, as is normal in business, or whether they genuinely liked me, I cannot say. But, whatever their reasons for being nice to me, it felt good.

One of the suppliers of the bulk cosmetic products I particularly enjoyed visiting was a company called IT & C, based in St Leonards on the outskirts of Hastings, located on the south coast of England. The company manufactured the bulk product before shipping it to the fillers located in various parts of the UK, transporting the material in one tonne containers called pallecons. It was only a small company, and I got on well with the two Directors, Jason and Oliver, as well as the Plant Chemist, Ben Alsop. Being a small business, and keeping it in the family, as they say. Jason and Oliver even had their personable wives Lynne and Sheila working in the offices.

Normally, I stayed in Hastings, and as Oliver lived quite close to the hotel, he would often insist on having a few drinks with me in a pub called the FILO, an acronym for, *First in Last Out*. At the back of the pub, the FILO possessed its very own micro-brewery. I discovered the FILO's own beers extremely agreeable indeed, consequently, during my many visits to the region, sampling a fair amount of their alcoholic beverages on offer, eventually sampling all the varieties of beers on sale, their eclectic variety of flavours, all proving amenable to my pallet. Oliver and I spent many a convivial evening in the salubrious surroundings sampling the beers on offer. During one of our many discussions, I discovered Oliver had a claim to fame. Apparently, for a brief spell of a few weeks, he had once played the drums in Cat Steven's backing group after the regular drummer had been taken ill. Oliver had also once been a fireman, but had received severe injuries on one call out, forcing him to retire. He received a substantial compensation package which helped him live fairly comfortably through his retirement; the package also provided him with the finance to buy into IT & C.

I later discovered from Ben Alsop, the Chief Chemist, Oliver had his fingers in many pies, and he seemed, well at least from Ben's account anyway, to be St Leonard's equivalent to Del Boy, or Arthur Daley. Ben recounted tales of loads of toys turning up at the site, apparently destined for the many arcade machines which proliferated the nearby seaside holiday resort. In addition, every morning, the office computer appeared to be inundated with multitudes of e-mails from various businesses, mostly addressed to Oliver. According to Ben, the e-mails certainly had nothing to do with cosmetic manufacturing and production. I liked Oliver immensely. He had certain character, a *bon vivant*, possessing a certain *joie de vivre*.

IT & C had been asked to manufacture one of SDM'S popular, regular, products, called Red-Hot Sauna Face Pack, normally supplied by one of the bigger bulk manufacturers. The product contained a material called molecular or micro sieve. The fine beads after coming into contact with the moisture on the skin, generated an exothermic reaction. When spread on the face, the moisture on the skin, caused the Red-Hot Sauna to heat up, and in the process, cleanse the pores. Well, that was the theory anyway. The Red-Hot Sauna could prove difficult to manufacture and so for the first batch, I travelled to St Leonards in order to give assistance and advice.

The initial stages went well. The process required the product to be heated and thoroughly mixed for a few hours to completely disperse the micro sieve throughout the bulk. But before proceeding any further, slides had to be taken to check complete dispersion had taken place. The slides would indicate if the first part of the process had mixed sufficiently before then proceeding onto the next stage. Get it wrong, and the results could prove disastrous, much like a recipe when baking a cake.

With the initial stages of the process well under way and quietly mixing in, it would be a few hours before carrying on with the next part of the procedure. With lunch time approaching, Jason suggested we travel to the FILO for a meal. So, Jason, Oliver, Ben and I toddled off to the pub for lunch. All paid for courtesy of IT & C. In addition to supplying a range of excellent beers, the FILO, also provided wonderful meals, with a fair selection on the extensive menu. It turned out to be a good lunch, very enjoyable, with much discussion and banter between the four of us. Before realizing it, a few hours had passed by, and I thought we should get back to the small plant to proceed to the next stage in the manufacturing of the Red-Hot Sauna.

In the instructions, it stipulated a time for the first stage, which had by this time, had been exceeded by almost thirty minutes. Ben had an extremely keen assistant, Keith. Keith went according to the specified time written into the procedure and carried on with the next stage, adding more ingredients. Unfortunately, he did not realize that he should first check the dispersion before proceeding any further. By the time we arrived back at the site, we discovered the next part of the process to be well underway. I had a quick check of the dispersion. Unfortunately, the micro sieve had not blended sufficiently in the first part of the process. Consequently, the product now looked like extremely thick, lumpy porridge instead of smooth yoghurt. I could not blame Keith, as I felt responsible. I should have been present all the time instead of swanning off to the FILO in hedonistic fashion. Ben and I spent the next few hours trying to get the product back to what it should be like. We could not throw it away and start again, the amount of material used so far had cost thousands of pounds, and as part of the business contract all paid for up front by SDM, I could not explain away using twice the amount of material as well as the time constraint. Besides, anyway, we did not have sufficient material to start again. Ben and I worked well into the evening jigging about with the formulation trying to achieve the correct consistency.

Finally, at nine that evening, we called it a day, deciding to carry on early the next morning. The next morning, we both turned up at seven at the facility and continued tweaking the mix. Both of us considering our futures, believing the job centre beckoned. I can tell the reader my stomach was churning over all the time. Finally, at lunchtime we managed to get the correct consistency in the product. Ben and I both felt drained both mentally and physically, but also immensely relieved. That afternoon, I had to drive back to Swansea, and being a normal Friday, I did not get back until late.

I never informed Harry Chivers or anyone else working at SDM, about the near disaster with the Red-Hot Sauna. I also knew full well, the guys from IT & C would also stay quiet about the whole episode. It would remain our secret; until now, that is.

Another interesting chemist from another company based in Wolverhampton told how, in order to supplement his income, he had acquired a part time job as an embalmer working for one of the local undertakers. Often explaining to me in great, gory detail the embalming procedure and how he took great delight in making young apprentices faint during the process. He would have had great fun teaching me.

Apart from the suppliers, I did not enjoy working for

SDM, but needed the money. An incident finally occurred which persuaded me I should leave the company, the circumstances of which, I will now relate.

I had to travel to one of the fillers based near Shrewsbury. They had quite a number of reels of laminate, which had been damaged during transit. They need SDM to sort out which laminates could be used, and which had to be rejected: a task which fell to me. The company had almost thirty pallets loaded with reels of laminate and which had to be inspected. I knew the job would take quite a while. I left my new digs in Swansea at five thirty in the morning, travelling up to Shropshire, arriving there around nine. The company had sent me an e-mail, listing the laminates and products etc. but I left it in the office, after all, they would have their own list. Knowing the task in front of me, was like sorting the wheat from the chaff, I started almost immediately after drinking a cup of coffee.

I didn't even have a normal lunch hour, instead taking twenty minutes. By three in the afternoon, I knew I could not possibly finish at a reasonable time. I phoned Harry, requesting I only work for a couple of hours more and book into a local hotel for the night and carry on the next morning. He told me, without any ambiguity, he wanted me back in the office at nine the next morning.

So, I carried on with the task until five thirty, until I felt I had inspected everything that needed to be looked at. The Quality Manager for the company, Russell, had left to attend his son's school concert. I asked his deputy if everything had been inspected and sorted. He came back and told me that the job was complete. All the pallets of laminates had been inspected. However, he did not have the list, but he felt certain there was nothing left to inspect.

Thus satisfied, at five thirty in the afternoon, I headed back to Swansea. It proved to be a cold wet miserable winter's night and the journey seemed to take forever, with wet, busy roads and oncoming headlights roads. I eventually arrived in the house at nine-thirty, that night, completely exhausted. I had been up since five that morning, with only one twenty-minute break for lunch.

I arrived in work at eight the next day and went straight to my office. At nine-thirty, Harry called me into the office, and told me straight 'You left early yesterday, there was one pallet of laminates left to inspect which you overlooked!'

'S'cuse me?' I asked, extremely bemused, and irritated. 'Left early?'

'Yes!' he retorted.

'Well they told me I had inspected all the pallets.'

'Did you check your list?' he enquired.

I told him I had left the list in the office, but as the company sent the list in the first place they should know what had to be inspected.

I had a written warning for only inspecting twenty-nine pallets out of the thirty and not working for an extra thirty minutes, only working a sixteen-hour day instead of sixteen and a half; and with no overtime payment. Perhaps a bonus of thousand pounds at Christmas for all the extra hours in a year. This was a return to Victorian times and conditions. Within a few months, I left the company, completely disillusioned. At least, with previous employers, my services had been appreciated.

'All change!' That expression has particular relevance, as the reader will shortly discover.

CHAPTER 22

After ending my employment with SDM, as a short-term stop gap, I had a brief spell working, *on the buses*. God, I love those three words, they bring back such evocative memories of Reg Varney and the TV sitcom from the late sixties and seventies bearing that name. This bizarre change in career direction came about following a discussion with one of Christine's female friends who suggested I might enjoy and benefit from the change in career direction. For the previous eighteen months, it appeared her partner had been driving for a major national bus company. The company provided most of the bus services in the Swansea area.

Apparently, her partner thought driving for the company a wonderful occupation and thoroughly enjoyable and would advise anyone who enjoyed driving, to take it up. So, whilst still employed by SDM, taking up the recommendation, I applied for a driver's position with the transport company. Unfortunately, during the intensive interview program, I failed the medical due to extremely high blood pressure. Obviously, an indication and warning the detrimental toll my position within SDM appeared to be taking on my health, providing me with yet another good excuse to resign. My body was telling me I should quit my job with the cosmetics company before it quit me.

Following several visits to my new GP in Swansea, I received medication to reduce my blood pressure; medication which I still take to this day. After what I considered to be a reasonable amount of time for the medication to kick in, I re-applied for a position with the bus company.

This time, my second application proved more successful and I passed the medical. My blood pressure now under control due to the medication, assisted by a slight alteration in my life style, such as reducing caffeine and alcohol intake, together with reducing my fat consumption. Within five weeks I had finished with SDM and commenced working for the bus company. This time with no regrets, in complete contrast to the situation with my previous employer, Chemkiln, a company I relished working for, a job I prized and a job which Bridget insisted I leave. Even leaving my employment with Hyperwaste, my employer prior to Chemkiln had been tinged with regret. For despite Nathan Edwards being extremely tight-fisted when it came to award any pay increases, he respected the work I did on behalf of the company. SDM displayed no such gratitude, respect or even a modicum of appreciation.

During those first few weeks working for the bus company, along with three other new starters, I received driving lessons, driving a forty-foot bus to obtain the necessary PCV licence. We all took turns at driving the coach under the supervision of Bruce Kayne, one of the company's three driving instructors. Any misconceptions which I possessed at the time concerning my innate driving abilities quickly became dispelled under Bruce's tutelage and instruction; swiftly bringing me down to earth and reality with a bump. During a previous incarnation, Bruce had once been a sergeant in the Royal Marines. Consequently, he displayed very few inhibitions when it came to show his displeasure at any mistakes perpetrated by myself or any of my fellow pupils during the driving lessons, often displaying bouts of irritability and extreme annoyance, punctuated by extremely colourful, profane language.

'You stupid fucking bastard use your fucking mirrors, look left and right; tick tock.'

While negotiating roundabouts, comments such as.

'The gate is closed, you can go, you fucking twat.

Call yourself a driver, you are a fucking moron. My ninety-year old grandmother can drive better than you. Use your gears; you fucking dickhead.'

Bruce exhibited no favouritism, frequently using these terms of endearment to every one of us fledgling drivers during the driving lessons. I must confess as well as improving my coach driving skills, my use of Anglo-Saxon also developed in leaps and bounds as I acquired Bruce's habit following, his driving instructions. As a result of my close contact with Bruce, I experienced great difficulty in speaking without inserting a considerable number of expletives during any normal conversations. It reminded me of my days as a young, naïve student working in the steelworks during the summer holidays, experiencing the same problem then in curbing my language. The people I worked alongside at that time using the same colourful language causing the effect to rub off on me.

One of my fellow pupils, a Londoner called Kevin, did nothing to help during my bouts at driving, often adding comments alongside Bruce's chirping in with his distinctive London accent.

'Look out for that car. There's a junction ahead slow down, change gear. Don't forget to use your wing mirrors; tick tock.'

I wouldn't have minded, if Kevin had been an exemplary pupil, but his driving skills appeared to be worse than mine. My simple reasoning behind this assessment is based purely upon the number of expletives and terms of endearment Bruce imparted while attempting to instruct Kevin. During his driving periods, I guessed Kevin received twice as many 'fucks' as I did. Frequently during Kevin's unwanted instructions and interference, whilst I drove, Bruce would interrupt and yell at Kevin to, 'Shut the fuck up. I am the instructor. Not you!'

Also informing the learner driver who happened to

be receiving Kevin's unwanted advice at the time, 'You can get the stupid fucker back when he is driving.'

Following Bruce's angry outbursts, Kevin would remain silent for a while, before resuming where he left off with his unwanted, unsolicited advice. There are more exploits and stories concerning Kevin to be imparted later.

Learning to drive a forty-foot coach proved to be a long, long way from chemistry and chemical engineering, and as if acquiring the necessary skills to manage the huge beast on a fine day tended to be difficult, assimilating those skills during a heavy snow storm proved nigh on impossible. One Tuesday morning, my turn arrived at driving the leviathan. A forty-foot long coach is not too bad, however when initially learning the requisite driving skills, it does seem like handling a leviathan. During my lesson the snow began to fall. Slowly at first, then the snowflakes began to increase in size and numbers. Soon a thin layer of snow began accumulating on the road, with no signs of abating. Even though it was a chilly day, the sweat began running down my back and I sensed Bruce becoming agitated. Suddenly he made an executive decision,

'Head back to the depot on the motorway and turn off at the Llangyfelach junction. We'll knock it on the fucking head today.'

The coach began to slither and slide on the icy, snow covered roads. I believed Bruce would take over the driving duties. Not a bit of it, instead he told me to stay calm, take it easy and not to panic. The last instruction came a bit too late. For, at that moment, I was in full panic mode.

The bus careered towards oncoming vehicles as they also slid on the icy roads, thankfully missing them by inches. The last stage of the journey meant negotiating a steep incline. Bruce ordered me to step on the gas to ensure the bus made it up the slope. I passed vehicles which had come to a stop after being unable to gain traction. With the snow still falling heavily, in fact heavier than twenty minutes previous, I guided the vehicle into the depot, slithering it to a halt. I felt quite proud of myself getting the bus back to the depot without any mishaps and without incurring even the slightest scratch, or dent, even receiving rare, albeit, brief praise from Bruce. 'Well done,' being the full summation of his praise, and without expletives, although, the episode probably did nothing to alleviate my blood pressure.

Thankfully, I did not have to endure Bruce's insults and haranguing for too long, eventually, after a few weeks, obtaining the desired full PCV licence.

Driving the commuter buses proved much easier than handling the huge coach. Firstly, the buses tended to be much shorter in length. Secondly, the commuter buses, for the most part, possessed automatic gears, in stark contrast to the huge, ancient coach which had manual transmission and was extremely difficult to engage.

For my first week I should have been accompanied by a mentor, an experienced driver who would 'show me ropes.' Unfortunately, due to an acute shortage of drivers I was told by Alan, one of the senior managers, simply to hop on the buses which he gave me list of and told to become acquainted with the various routes. Therefore, my first week involved simply jumping on and off various buses and journeying around Swansea. Mainly, the Mumbles and West Routes, although Alan did give me some other routes to learn.

With so many routes to learn, my knowledge of the roads tended to be extremely rudimentary, so imagine my surprise when turning up for work on the second week, Alan asked me which routes I knew best, when I told him the West Cross and Mumbles. He instructed me to take one of the buses out on my own: the number three bus to Mumbles. I was a bag of nerves. Unfortunately, due to my inexperience, that bus ended up running late and as I had to do a few trips, the time deficit increased. Along with driving, the driver had to issue passenger tickets when they boarded. With the increasing demands to cut costs, no clippies worked on the buses as in days of old.

Unfortunately, after that day, I was put straight onto the Mumbles and West Routes, receiving no mentoring whatsoever. Drivers who started after me had the luxury of being accompanied by an experienced driver for at least a month.

On my second day, I turned up early for my shift at the Quadrant, the main bus terminal in Swansea. The Controller asked me if I could take one of the buses out for him as one of the drivers had not turned in and he was short staffed. I had only been on the route for one journey and not au fait with the roads. As result I took a wrong turning and had to be guided by one of the regular passengers who directed me. One of the regular drivers warned me about some passengers. He had once been on a route he did not know, and a passenger said he would guide him. The double decker ended up going down narrow country lanes. When the passenger suddenly told him to stop as he was outside his house and thanked the driver for bringing him home, situated in a cul de sac, the passenger immediately alighting from the bus. The driver had to reverse the double decker for about half a mile back up the lane before he could safely turn it around. 'I ended up in a bath of sweat and a jabbering wreck by the time I could turn the bus.' the old driver informed me.

I also heard the story of one driver who missed the turn off on the road to the Gower peninsula. He ended up driving to Worms Head, a round-trip of seventy miles. The perils of driving the buses around Swansea.

CHAPTER 23

When surveys are carried out involving the public concerning people in the service sector, and the question asked, 'Who you would consider to be the rudest people in public service industry?' Invariably, the answer always appears to be, 'Bus Drivers.' I am not going to dispel this belief or even argue with the statistics, for quite a few of the drivers I worked alongside made no bones about the fact they hated the passengers, those feelings often being reciprocated by the commuters towards the same drivers. Many of the drivers, particularly the morose, curmudgeonly ones, while in the staff canteen, often declared their feelings concerning the passengers, habitually pronouncing, 'I love driving the buses, pity we have to carry fucking passengers.' I must say, some of those drivers appeared to hate the passengers with a vengeance.

Also, many people do not like using public transport. I recall an attractive woman getting off the bus. While waiting for the bus to stop she confided in me.

'I knew there was a reason I don't like using the buses. It's the other passengers.'

She then continued.

'Did you notice the disgusting, drunken man who got off a few stops back?'

I told her I hadn't noticed him particularly.

'Well, he gave me a big slobbering kiss before he got off.' The disgust was evident in her face and demeanour. Such are the pitfalls of using public transport. You can well understand why some people are averse to using the transport system.

I remember one of my colleagues, Craig, a young driver in his twenties, would often turn up in the canteen with some story or other about the day's events. Once he ended up arguing with some students from the University halls of residence at Hendrefoilen on the outskirts of the city. The argument became so heated he had to drive away leaving the passengers behind. The students displaying their anger by throwing assorted items, such as coke bottles and half consumed food containers at the bus, while Craig drove his vehicle down the road at breakneck speed endeavouring to get away from his pursuing adversaries as fast as possible.

On another occasion, the company intended taking the buses off one of the rural routes on the Gower peninsular. Craig had one of the regular elderly passengers protesting vehemently about the removal of the buses from the route asking how he was expected to travel in the future. Craig sat calmly, thoroughly bored, perched on his seat frequently replying, 'I don't care, I just don't care.'

Craig tended to be extremely bothered and concerned about the plight of passengers an extremely caring person… I think not. The company, after holding discussions with the city council, decided to rescind their decision. Probably after being promised some monetary incentive, they kept providing buses on that route. Craig had the misfortune to meet the passenger once again. The old pensioner, exhibiting a huge smile decided to wind Craig up. 'Power to the people, we had the last laugh Sonny Jim, thought you had seen the last of us eh?' Another time Craig walked into the canteen 'I've just had the fright of my life I just had Kevin driving hurtling towards me on a narrow lane driving a huge coach. I don't think he even saw me and appeared to be in a dream. I nearly smashed my bus up. Thankfully, I only had a slight bump on the roadside hedge. Just hope I can get away with it?'

Christmas time is a time of joviality and good will towards all men. Not this Christmas, the company had laid on a special bus tour trip around Swansea for the little kiddies to see the sights and the many Christmas lights festooned around the city. Unfortunately, although it must be stressed, unintentionally, the company put one of the most miserable drivers on the tour. Dressed in his Santa outfit, the driver made it abundantly clear he did not want to be transporting a load of screaming kids around the Welsh City, the week before Christmas. Instead of gently handing out the Christmas boxes to the excited children he threw them down the aisle of the bus often hitting the kids on various parts of their bodies.

Any boxes which inadvertently landed on the floor inevitably felt the toe of Santa's boot before hurtling inside the bus at great speed. Why throw the gifts at the kids when it is easier to kick them at the little darlings? Any questions from the excited children usually received some sarcastic reply. 'How the hell should I know? Do think I am the font of all knowledge?' There were a lot of unhappy children at the end of the tour. After receiving numerous complaints, the driver received his marching orders from the company. Talk about 'Bad Santa.'

Kevin, my fellow learner driver gained a reputation in the company for all sorts of reasons. One day, he boarded a bus parked up in the quadrant to do his route and began altering the destination board. A lone passenger sitting on the bus asked, 'I thought this bus was going to Killay?' Kevin told the woman plainly it was destined for the Mumbles; a totally different route. 'Isn't this bus going to Killay?'

Kevin became quite dogmatic and belligerent. Despite the passenger pointing out, that bus always went to Killay. After a heated discussion, the passenger got off the vehicle in a huff, So Kevin took the bus along the sea front. After a while, the controller came in the Radio. 'Kevin Where the hell are you?' to which he replied, 'Just coming into West Cross by Mumbles.'

The controller became irate. 'What the hell are you doing there? You are supposed to be going to Killay!' That was Kevin for you, for some inexplicable reason, he had taken it into his head to take the bus on the wrong route, in completely the wrong direction.

Another time the company in Plymouth had an acute shortage of drivers and asked the company for drivers to alleviate the manpower shortage. Kevin volunteered to go to Plymouth. He spent a week route learning, but after that week decided he could not learn the routes and returned to Swansea, much to the management's annoyance. They had to pay him for the week plus living allowance. I happened to be in the office when Kevin received his pay. The manager sarcastically commented 'Not bad for doing nothing.' Unsurprisingly, shortly afterwards, the company terminated his employment.

The last I heard about Kevin, he left to drive for another company travelling on the continent but true to form travelled to the wrong Lille on the continent, with a bus load of holiday makers after punching the wrong Lille into his sat nav. Instead of Lille in Belgium, he headed for Lille in France. That particular episode even made the national news and headlines.

The company insisted on using their old, clapped out buses on the Townhill route, primarily because of vandalism from the youths who lived in that council estate. Unfortunately, with the estate being perched on the top of a mountain, the roads tended to be quite steep. There was one road called Mount Pleasant which the buses had to negotiate. Often the old, clapped out buses, when fully laden with passengers, could not cope with the steep incline. On many occasions, some of the healthier passengers had to get off the bus to lighten the load and walk alongside. Just like the pioneers with the wagon trains travelling west on their journey to the American frontier.

With that Townhill route, I remember working an evening shift when one of the new young drivers came on the radio, a hint of desperation evident in his voice. 'I'm doing the Townhill route and every time I go past Leakers on the Mount Pleasant road, the young kids sitting outside the café go to the back of the bus, open the panel and push the emergency stop button, causing the engine to cut out.'

Over the radio, the controller gave him some advice. 'Just speed up when you are passing them!'

The young driver came back with his reply, the croaking in his voice palpable, indicating his despondency and the apparent hopelessness of his situation.

'I can't, they do it when I am travelling up the hill and they can walk faster than the bus.' At that moment, I had to stop my bus, I was laughing so much. I suspect there were a few other drivers working that night doing the same thing. Still the company kept insisting on using the clapped-out buses on that route.

One driver who began working for the company, shortly after me, somehow acquired the name Wishee Washee' His real name being Peter. When I learnt his nickname, I asked one of my colleagues why he had been given the pantomime name. My fellow driver began recounting the tale, a huge smirk on his face. He began explaining about the automatic bus wash at the depot. I must explain, every other day, the buses had to go through the bus wash. The colour sticker on the front of the bus signifying if the bus had to go through; Red one day blue another.

During his first week, Peter took his bus through for the first time. Unfortunately, he managed to put the wheels on the nearside of the bus outside of the guide rails. Before he noticed it, the brushes began moving alongside the side of the bus. Instead of the brushes, gently washing the bodywork, the metal prongs, holding the huge brushes in place, began making a sickening noise as they scraped and grated alongside the body of the vehicle, metal against metal.

Peter tried to reverse the bus out of this predicament, all to no avail. The bus remained jammed solid, as the metal pushed the side of the bus, forcing the wheels against the outside of the guide rails. The bus would just not move. Inexorably, the metal slowly gouged through the side panels of the bus. Unfortunately, no one was around to push the emergency stop button of the bus wash. And Peter could not get out of the bus, the offside brushes of the machine barring his exit. He just had to sit there while the huge machine inexorably ripped a large slit along the side of his vehicle. Once the brushes had arrived at the back of the bus, Peter managed to quickly get out and push the stop button of the bus wash. By now, the bus had an unhindered slit running from the front to the back, as if a giant tin opener had been operating on it. The final cost for repairing the bus was a whopping £10,000. Peter never got fired but he did receive a written warning plus his new nickname into the bargain.

I know I have previously written about ghostly stories, I am of course referring to the episode in the Brockett Arms, situated in Ayot St Lawrence. But I had another experience, although being the sceptic that I am, I'm more inclined to believe it was more a case of auto-electrical malfunction than a poltergeist.

The company possessed one bus, ironically painted black which gave it the distinct appearance of being a giant mobile coffin. One evening while driving this bus back to the depot at the end of the day, at about twelve midnight, the bell intermittently kept going off, although, there were no passengers on board. This ringing of the bell kept occurring every few minutes. Also, the lights inside kept dimming and on one occasion the engine cut out for some inexplicable reason.

Upon getting back to the depot to pay in all the money I had taken that day, I told one of the other regular drivers about my nerve-wracking experience, when I informed him of the bus number, and its livery, he casually remarked. 'Oh, that's the haunted bus. Haven't you driven it before?' I informed him that was my first time at driving it. It seemed the manifestations I experienced had happened to numerous other drivers. I never drove that bus ever again, as shortly afterwards the company sold it to another firm. I am personally convinced the bus had electrical problems. Having said that, the bus had been checked out numerous times concerning the problems. I also put in a maintenance request, yet nothing was ever found by the auto electricians. So, who knows the real reason?

I came across quite a few characters during my short sojourn working on the buses. There was another character Ray. Ray hailed from the West Midlands, and spoke with a distinctive Brummy accent, but he possessed a wonderful sense of humour. Like Craig, one of the drivers I mentioned, he would often turn up in the canteen with one story or another. A couple of tie stories I can still recall and which I still find highly amusing to this day.

This day, he had been driving the X21 to a village called Clydach, located on the suburbs of Swansea. Now on the route every few hours the bus had to make a detour to Waverley Close, a cul-de-sac with a large residential home for the elderly situated at the end of it. The bus had to turn around and return to the main road and resume the normal route on the schedule. For this particular schedule, Ray did not have to take his bus up Waverley Close and carried driving up the road. Suddenly there was a frantic ringing of the bell and a raised voice coming from a curmudgeonly old gentleman with a walking stick sitting in the middle of the bus.

'Why haven't you gone up Waverley Close? This bus always goes up Waverley Close?' He vociferously complained to Ray. Ray stopped the bus and an argument ensued between himself and Clydach's answer to Victor Meldrew.

Ray showed him the schedule and indicated to the belligerent old man that he did not have to go up Waverley Close. The old man responded angrily.

'I have been catching this bus for 5 years and the 4:15, X21 from Clydach always goes up Waverley Close!' Ray answered in his calm, Brummy accent, immediately realising the pensioner's mistake. Entirely comprehending the situation and knowing full well, he was in the right.

'I am not disputing the 4:15 bus goes up Waverly Close, but this is the 3:15. You didn't put your watch back yesterday at the end of British Summer Time, did you?'

The angry, old age pensioner appeared to be at a loss for words, suddenly realizing his mistake and how he had made a complete idiot and total prat of himself.

'Huh! You think you are right clever sod, don't you?' With that, in a huff, he exited the stationary bus now with a longer walk than he had envisaged, in front of him.

On another occasion, travelling the same route, Ray recounted yet another tale about the X21. Approaching the end of his route at the top of a place called Craig Cefn Parc, he suddenly encountered a highly coiffure, highly made up, middle aged woman wearing an expensive fur coat driving a top of the range BMW. *All fur coat and no knickers,* as they say in the valleys. That section of the road narrowed considerably, permitting the passage of only one vehicle at a time, there was certainly no room to allow the passage of a bus and a rather large, expensive BMW.

The woman considered it far easier for Ray to reverse his long bus than for her to reverse her BMW. She absolutely refused to give way, considering Ray should reverse his bus and not her. Ray related of how he put the handbrake on and pulled out the Mirror Newspaper and casually began reading the day's news. Meanwhile, behind his bus, a Funeral cortege had joined the main road from a side road and slowly approached the back of the vehicle. It headed towards the small graveyard at the top of the hill. Despite her demure appearance, the woman began uttering profanities at Ray telling him to reverse his *fucking bus*. She appeared unrelenting. Poking her head out of the driver's window, she shouted at Ray, 'I am not going to move, you will have to reverse your fucking bus!'

Ray casually replied in his Brummy accent, 'Fine; you can tell that to the funeral procession coming behind me!'

The woman looked around the side of the bus and could plainly see Ray was not lying or exaggerating. Indeed, a large funeral cortege was slowly heading up the road towards the back of the bus. So, with a huff, and uttering a multitude of new profanities, the fur clad woman reluctantly reversed her BMW up the narrow country road, allowing Ray to proceed on his journey. Such were the joys of being on the buses.

Working on the buses proved to be a different and interesting experience. Driving the last bus from the centre of Swansea to the outskirts, with the passengers either fighting, arguing or suddenly bursting into song, some drivers referred to the last bus as 'The singing bus.' Or, sometimes, there were young teenage girls, performing sexual acts on their boyfriends at the back of a bus. Not an uncommon occurrence.

Not long afterwards, I saw an advert submitted by a local employment agency. They were looking for chemists/chemical engineers to work at the steel works in Port Talbot. 'All change once again.'

CHAPTER 24

As the saying would have it, what goes around, comes around. I refer specifically to this snippet of home spun philosophy, because the first job I had in the late sixties, and at the beginning of the early seventies, managed to be in a steel works located alongside, and dwarfing my home town of Ebbw Vale. I worked in the aforementioned large industrial facility during the summer holidays, primarily to earn some extra money while still studying at Swansea. Years later, during the noughties of the new millennium, the last full-time job I acquired also transpired to be in another steelworks; this time, with a slight difference, not the steelworks of Ebbw Vale, but a huge, much larger steelworks located on the South Wales coast, and near to where I lived in Swansea with my wife Christine.

During my brief, but illuminating spell working on the buses, I enrolled with an employment agency for engineers and industrial chemists. One day, I opened an e-mail sent by the agency, which suggested I should apply for a temporary job working for the nearby steel plant. I sent off my updated CV to the Energy Department of the steelworks, which had advertised the vacancy. A brief time later, the Personnel Department sent me a letter asking me to attend an interview, specifying I bring a short Power Point presentation saved on a memory stick or disk. The memory device to contain a presentation concerning water and effluent treatment, together with a section devoted to the production of de-ionized, high purity water for power plant turbines.

The letter also stipulated the short presentation should last approximately fifteen to twenty minutes. How times had changed from my previous job interviews, when I simply turned up; now presentations, memory sticks and computers came into play. I remained undaunted at giving a presentation. The only major obstacle, how to keep the talk down to within the stipulated twenty minutes, yet still manage to put over all of the salient points concerning the subject matter. The presentation covered quite a wide and eclectic range of topics.

A quarter of an hour seemed insufficient time in which to do it justice. Firstly, it meant giving a rehearsal presentation to Christine at home, armed with a stop-watch, attempting to get the timing right, keeping the presentation to within the restricted period. After quite a few practice runs, and many amendments, I finally felt comfortable. Managing to crop the presentation to just under twenty minutes, yet still managing to get over the salient points. The salient points, I felt, would give an indication concerning my knowledge and comprehension of the areas which I considered required mentioning.

All prepared, I turned up for the interview at the main offices of the gigantic steelworks. Phil Evans, the Energy Manager, presided over the interview, assisted by Mike Hicks from the Personnel Department. Upon my arrival, after first introducing themselves, Phil Evans immediately set up a laptop computer, then opened a large collapsible screen, ready for my presentation. He asked for my memory stick, plugged it into the USB port of the laptop, and then instructed me to begin. For some inexplicable reason, while giving my presentation, I felt completely relaxed, giving my talk without any hiccups, and thankfully, within the stipulated time. Both Phil Evans and Mike Hicks asked me a few questions about my work experience, and whether I could work under pressure. That was it.

I must have given quite a reasonable presentation, for a few days later I accessed a voice message on my home landline. The recorded message came from Phil Evans, offering me the job, and asking when I could start, stipulating that I would not be working for the steel company directly, but through my employment agency. The term of the employment would be for a minimum of eighteen months, but the contract would be open ended, and could possibly last for years. He also informed me, my line manager, and head of department would be a person called Alan Prescott.

Despite the job not being on a permanent contract, I decided to accept the position, feeling I needed to return to my vocation before finally retiring. Working on the buses had been interesting, but it was not really me. The weird shift patterns, together with the low wages, all augmented the negative aspects of the job. I immediately phoned Phil Evans, accepting his offer, telling him I could start within a month, after giving my notice.

The remaining weeks working on the buses seemed interminable, but at last, the termination day arrived, and I finally left the bus depot for the last time. Unfortunately, on the weekend prior to commencing my new job, Christine took it into her head to give me a much-needed haircut, in preparation for my new position. Alas, she shaved too much off a small section of my head, which made me look weird. Almost like an old, grey-haired Mohican. The only answer to this predicament was to attempt to even everything out on my head, by giving me a 'number one' haircut all over. Completely eradicating the small amount of my remaining hair. Adopting hairdresser's parlance, *cutting it right down to the wood.* Hence, I turned up for work on the first Monday morning, looking like an old aged skinhead.

In fact, my appearance altered so much, Mike Hicks, the Personnel Manager, did not even recognize me, walking straight past me in the reception area. When I approached him, he gave me a startled, odd, strange look, whilst muttering something to the effect, 'I looked nothing like the person he had interviewed weeks earlier.' He seemed to have the impression I looked like some sort of thug or hoodlum. In truth, he had hit the nail on the head. For, I must confess, I did look like an aged thug or skinhead; courtesy of my wife. Eventually, he finally accepted I was indeed the person he had interviewed weeks earlier, ultimately contacting Phil Evans to come and collect me.

For those of you who have never visited a steelworks, I must first explain, they are huge facilities. The steelworks at Ebbw Vale could be described as large. However, this steelworks belonged in a different league, falling into the 'ginormous' category. The gargantuan site completely dwarfed the old site at Ebbw Vale.

I spent those first weeks becoming acquainted with the Energy Department. The personnel in the department, together with my domain, seemed to be everywhere around the site. My mentor during the first weeks was a chap named Jeff Edwards. Following Phil Evan's promotion, Jeff had been 'minding the shop.' Jeff, being a Mechanical Engineer, not a Chemical Engineer, felt uncomfortable in the position, not being one hundred percent confident with his chemical knowledge.

Once again, I embarked upon a steep learning curve, having to become acquainted with all of the internal Computer systems; TEMS (Total Engineering Managements System, and KPI's (Key Performance Indicators), accounting systems and learning the intricacies of the department. Additionally, it meant getting to know other personnel, and their functions within the departmental structure.

The extent of my duties proved to be extensive, and it appeared I had to look after the water supply and the quality coming into the plant, together with the effluent system feeding into the sea, as well as the de-ionized water systems feeding the plant and power plant, which provided electricity to the immense facility.

My duties and responsibilities increased immensely, differing markedly from my time working in the old Ebbw Vale steelworks. Then, at that time, in the late sixties and early seventies, I had been a lowly student labourer, only working to acquire some money while pursuing my studies. Now, my responsibilities appeared quite daunting, involving virtually all the water feeding the works and the effluent discharging into the Bristol Channel. I was also responsible for a large reservoir, something new in my experience. For those readers who have been fortunate enough to view the classic film '*The Dam Busters.*', the film makes it clear the importance which water plays in the production and manufacture of steel, and is virtually, the life blood of the industry. Well, at least I didn't have the fear of it being bombed by RAF 617 Squadron.

I discovered the extent of my duties also included looking after and maintaining sufficient stocks of sodium hydroxide and acids, used in the regeneration of the de-ionizers. If those units shut down, the financial implications for the facility could prove extremely expensive. I also had to put together a new contract for the companies looking after analysis of the water supplies, ensuring the cooling waters, especially the huge cooling towers, remained free from the legionella bacteria, as well as the air conditioning and water systems within the plant.

The contracts for analysing and maintaining the water systems had been in place for four years, and were due to expire within the year. My on-going job involved putting together a new, updated multi-million-pound contract for the water companies to submit tenders. The contract consisted of two main areas: the Heavy End, which comprised mostly of water systems for the two Blast Furnaces, and the power plant, looked after by a company called NALCO. The other end comprised of two huge cooling towers for the rolling mills, the steel making plant, pickling lines as well as an old Demin plant feeding that section of the facility. All this section was looked after by a company called Betz. I must admit, I found the task daunting, putting together a high-powered contract for the huge steelworks, a steelworks I had yet to become fully acquainted with. Talking to some of the guys in the department, there were areas of the plant they did not even know. To discover every nook and cranny of the immense works would take an individual years.

During my time working for other companies, whenever costs ran into a few thousand or even hundreds of pounds, then invariably a lot of questions tended to be asked by higher management. I quickly discovered working for a steel company proved to be in a completely different league. This realization first came about when the personnel responsible for the day to day running of the de-ionization units, which provided the ultra-pure water for the power plant, running the turbines for electricity generation, informed me one of the polishing units, which removed the final traces of elements and salts from the first stage of the deionization process, continually needed regeneration.

The regeneration process necessitated washing the mixed anionic and cationic resin with sulphuric acid and caustic soda. This particular polishing unit required three times the regeneration of the other polishers. An indication something appeared to be not quite right with the resin bed or the unit. I approached Alan Prescott and told him the unit needed to be looked at and overhauled, which would mean renewing the resin bed. He instructed me to get quotes for servicing the unit. The unit had not been looked at for years, possibly even decades.

Eventually, I chose a national contractor to perform the servicing. Their quote, being the most reasonable, coming in at around seventy thousand pounds. The task had to be performed during the summer shutdown period, when the power plant went through its annual maintenance programme. The servicing intended to last, about seven days, allowing plenty of time during the two-week shutdown period.

Unfortunately, after accessing the unit and removing the old spent resin, it became evident much more work had to be performed on the unit than had originally been planned. Firstly, the protective coating which lined the metal shell of the unit had started to break away and needed replacing. Secondly, the diffusers feeding the water, which also dispersed the regenerating acidic and alkaline solutions into the unit, had completely disintegrated and collapsed. No wonder the unit failed to perform properly.

What should have been a routine servicing, to be performed in plenty of time, now became a tremendous task, which had to be carried out, working around the clock, with time tight on getting the unit back up and running in ready for the start-up of the Power plant. Not only that, because of the additional work, the costs also increased from the original seventy thousand pounds, doubling to around one hundred and forty thousand pounds. I had to give an answer rapidly for the work to commence, ensuring completion before the power plant start-up. I approached Alan Prescott, who had a sharp intake of breath when I presented him with the quote from the company doing the servicing.

'I can't sign that.' he explained. 'You will have to ask Russell Davies, the Heavy End Director. I am not going to ask him. We are way over departmental budget as it is. You will have to go on your own.'

So, with a heavy heart, and a churning stomach, I headed for Russell's office. I had only met him a few times and expected a right ear bashing about the escalating costs. He had only recently been promoted to his position, hence the reason for Alan's reluctance in approaching him with the unwelcome news. Much to my surprise, Russell appeared quite calm and pragmatic about it. 'I had heard about the problem,' he informed me. 'It is just one of those unforeseen circumstances which transpire from time to time. As they say, "shit happens. Go and make out the purchase requisition order for the additional seventy thousand pounds, and I will sign it off.' The only person more surprised than I at Russell's calm, pragmatic attitude, proved to be Alan Prescott.

One of the other engineers within the department, and who had responsibility for the pipelines feeding the coal gas from the coke ovens to the blast furnaces decided to have them thoroughly cleaned. It would be the first time in twenty-five years. His costs came to a humungous three million pounds. As previously pointed out, the costs for manufacturing steel tends to be in a different league from that of other, smaller industries.

I referred earlier to RAF 617 Squadron, who almost brought the Third Reich's steel industry to a complete standstill. Well, I had the dubious reputation of being involved in almost bringing the colossal Welsh steelworks to a halt. One of the large de-ionizing plants used hydrochloric acid for regenerating the cationic resin. The resin absorbed metals from the incoming water. When saturated with the metals, the resin required flushing with hydrochloric acid to remove them, a process known as regeneration. After regeneration, the column of resin could be put back into service. There were a few of these units in the de-ionization plant. This particular plant did not supply water of such high-quality purity as the one supplying water for the power plant, but of sufficient quality to help produce excellent quality steel, a vital part of the process.

I continually monitored the levels of the acid for the cationic resin, and also the caustic soda used for the anionic resin. This particular week, I had ordered twenty tonnes of acid as normal from the supplier, a company called INEOS. Unfortunately, the next day, I had a call from their Customer Liaison Department, informing me the company had technical problems with their facility in Scotland. Consequently, the delivery would be delayed for an estimated twenty-four hours. At the time, I remained unconcerned, as the storage tank had a sufficient amount of acid for the time being.

However, I did become concerned later that day, when, once again, INEOS contacted me, informing me they had difficulty in resolving the technical manufacturing problem. The delivery would be delayed another forty-eight or possibly, seventy-two hours. Now I became concerned, having to think fast. The volume of hydrochloric acid in the storage tank, by now, approached the minimum safety level, with only one day or possibly two day's stock at the maximum. The only option, to obtain a slightly lower quality hydrochloric acid used in other departments or sister plants and supplied by another manufacturer other than INEOS.

Once again, I had to call on Russell's help. He had contacts in other sections, and able to call in a few favours. Otherwise, I could be held partly responsible for the shutdown of a huge steel plant and join the annals of RAF 617 squadron, in stopping steel production. Fortunately, after a few phone calls, and much cajoling, Russell, miraculously managed to acquire some hydrochloric acid from a steel plant in Newport, which had quite a large reserve of the acid.

I never thought birds, I refer to the feathered variety, would ever play a part in my working life, but they did during my time at the steel plant. Part of my duties included looking after the oil canals, where oils from the mills fed into, the viscous oil layering on the surface. One day, I had a frantic call from the supervisor of one of the fabricators carrying out work on the canals. Evidently, a bird had inadvertently got through the protective mesh, and landed in the oil. I arranged to meet the supervisor, who had, apparently, put the polluted bird into a cardboard box. I met him, as agreed. He immediately showed me the bird, which did appear to be quite distressed, and completely covered in oil.

'I have never seen hairy-arsed welders get so animated about a bird before. They are most concerned about it. You would never credit how grown men can get so distressed about a feathered creature,' the supervisor informed me.

I asked him what he expected me to do about the bird, being the Water and Effluent Engineer, and not a Vet. I managed to contact the RSPCA who immediately sent one of their officers around. After arriving on site, he took the bird away. The officer told me he felt confident he would be able to clean the creature and save it. The next day, the officer phoned, telling me the bird had been cleaned up, and appeared to be doing well. It would be released into the wild within a couple of weeks. At the earliest convenient time, I went to the contractor's cabin, to impart the good news. The supervisor had not been wrong. When I told the welding crew, they almost broke into tears. It just goes to show how even grown men can be affected by animals.

Another episode relating to birds, involved the large reservoir. We had a civil engineering consultancy firm who advised the company on the reservoir, making sure we conformed to The Reservoir's Act, the Energy Department being responsible for the large expanse of water. During one of his visits, the MD of the consultancy firm noticed some trees beginning to sprout along the grass embankment. He advised me to remove the trees before the roots took hold and began affecting the integrity of the concrete structure, which lay just beneath the earth. I had a chat with Phil Evans who had previously been responsible for looking after the reservoir. He also advised the trees to be taken down as soon as possible.

Within a week, I had arranged for a local landscaping firm to remove the trees. No sooner had the job been carried out, when I received an irate phone call from a local member of the RSPB. The local area's equivalent to Roy Cropper. He began berating me about the removal of the trees and how it would affect migrating birds and the places for them to nest. I must admit, I felt unsure of my ground. Should I look after the responsibility of the reservoir and the safety of people, who could be affected should a reservoir breach, or did I have to leave the foliage for the ornithological community? After an in-depth discussion, I apologised profusely and agreed to contact him should any further remedial work be required to be carried out to the reservoir banking. Thankfully, we ended our discussion amicably.

Working in the Energy Department certainly brought me into the realms of an eclectic range of issues.

CHAPTER 25

It did not take me long to discover the power and influence which the Energy Department wielded within the large steelworks. The various areas within the huge facility maintained their own effluent treatment plants, for which they held responsibility. For example, the BOS plant which converted iron into steel, maintained one of the largest effluent treatment facilities. The pickling plant, slab mills, blast furnaces, also had their own individual treatment plants. All these treatment plants fed into a giant reception tank for which the Energy department, when I say Energy Department, I mean myself and in the early days, assisted by Jeff Edwards, my mentor, held responsibility. The theory being all the treatment works within the site, kept their effluent within the permitted discharge limits. The huge reception tank simply regulated the flow of the neutralised effluent into the Bristol Channel.

Well, that was the theory. However, unfortunately, the BOS plant could not keep the pH of their effluent under control. The name BOS (Basic Oxygen Steelmaking) gives a clue. The effluent tended to be of high alkalinity. Consequently, the BOS department kept sending effluent, with extremely high, alkaline pH into the reception tank, which in turn affected the final discharge. Frequently, we had to divert the effluent from the large tank into a large lagoon, rather than discharge into the channel, contravening our discharge limits, giving us time to resolve the issue, before discharging into the channel. This breach would result in heavy fines, levied by the Environment Agency.

When I joined the company in the first decade of the millennium, during the noughties, the global world economy began approaching its zenith and booming. The demand for steel was unparalleled in previous history. The huge plant could just not supply enough steel. The evidence, made apparent, by the main railway between the Blast Furnaces and the BOS plant, which passed directly in front of the Energy Department offices. The relentless procession of torpedoes (torpedo being the name given to the containers holding the molten iron), pulled by train, containing hundreds of tonnes of molten iron, passing directly in front of the offices, on their way to being converted into steel at the BOS plant. The procession of torpedoes proved to be frightening, with the possibility that one of the torpedoes could accidently rupture, spilling its molten contents, the contents ultimately enveloping the office building in which I sat, along with others. The passing trains generated their relentless, hypnotic sound as their wheels passed over the expansion joints: repetitive sonic evidence of the prosperity in the world economy at that time.

The increased demand also meant highlighting the various weaknesses. The BOS plant had its failings in their treatment plant, the increased production exposing those deficiencies. When I first contacted the BOS Plant Manager to complain about the problems which the BOS Plant effluent caused, I detected a slight hint of nervousness and subservience in his voice. It took me aback somewhat, rarely, did I instil such fear and alarm in people. Then, perhaps his apprehension was not because of me, but at the thought of the possible consequences, should the huge steel facility receive bad publicity and resultant prosecution should we fail to contain this infringement, an infringement caused by the BOS plant.

After a few such episodes, a meeting was arranged to resolve this recurring problem. Eventually, it was decided to install extra acid storage for the BOS Plant effluent plant to assist in quickly bringing the pH down and under control. Also, Alan Prescott agreed the Energy department should also put a large acid tank by the huge effluent reception tank as an additional safeguard. The acid could then be fed into the large reception tank to bring the pH down.

Whenever I had cause to complain to other departments about their effluent discharge, I always detected the same fear and trepidation in their voices. Ah, the feeling of power! I am not a megalomaniac by nature, as the feeling of power might have gone straight to my head. But, with power, also comes responsibility. I remember one day I had to be at three separate meetings all at the same time. Crazy or what? Obviously, two of the meetings had to be shed.

All good things must come to an end, and my time working in the giant steelworks did just that, abruptly ending as quickly as it began. Previously, I mentioned the thriving world economy, and, of possibly being a slight jinx. After all, perhaps it is because of me, not the duplicitous, greedy bankers, that the economic bubble ultimately burst in 2008, when everything came crashing down, sending the world went into a deep recession. Consequently, I suffered, along with many others who lost their job in the steelworks along with other numerous industries and occupations.

Working through an agency and being one of the last to join the workforce, I ended up being one of the first to be made redundant, as the world demand for steel suddenly declined, and the world economy went into freefall. Both, the Blast Furnaces and BOS plant were put into ticking over, production ceasing for quite a while. The relentless movement of the torpedoes ceased, and they suddenly stopped passing in front of the Energy Department offices. The railway line went strangely and eerily silent. For some time, the plant began losing one million pounds a day. It all goes to show how quickly things can change.

The steelworks proved to be my last full-time job. I went into retirement, allowing me the opportunity to write and reminisce about my working career. While typing away on the keyboard, the writing experience proved to be cathartic, often bringing a smile to my now ageing face, remembering those people I once worked alongside. Some I liked, some I absolutely loathed, and some I have even forgotten about. Quite a few of those individuals, the ships that passed in the night, whom I once knew, have long since gone to *the spirit in the sky,* as sung about by Norman Greenbaum.

As a *Baby Boomer* born in Britain in the early fifties, I lived through the best period of the twentieth century with the swinging sixties, free love and the pill. More disposable income, reasonably affordable housing, no military conscription, no major war (apart from the Cold War), a National Health system providing free health care, plentiful employment and good wages, free university education, an acceptable retirement age, with a satisfactory state pension, enhanced by a private pension. We (the *Baby Boomers*), had it all, and I am thankful and eternally grateful for that. I am also aware and concerned future generations in the twenty first century will never have the same privileges and benefits we enjoyed. Their life will be far more daunting, tougher and demanding than ours. The *Baby Boomers* certainly had the best of it.

Finally, when all is taking into consideration, with my working career, I wouldn't change any of it and have certainly experienced life with all its adventures, along with its ups and downs. The people I encountered along the way, enhancing and enriching my particular, personal journey through life.

As I said in my first effort at putting pen to paper, or, to be more accurate, fingers to keyboard.

It is *Life's Rich Mix.*

~ THE END ~

www.ingramcontent.com/pod-product-compliance
Lightning Source LLC
Chambersburg PA
CBHW031612210526
45464CB00004B/1543